5G Enabled Secure Wireless Networks

Dushantha Nalin K. Jayakody
Kathiravan Srinivasan • Vishal Sharma
Editors

5G Enabled Secure Wireless Networks

Springer

Editors
Dushantha Nalin K. Jayakody
School of Computer Science and Robotic
National Research Tomsk
Polytechnic University
Tomsk, The Tomsk Area, Russia

School of Postgraduate Studies
Sri Lanka Technological University
Padukka, Sri Lanka

Vishal Sharma
Department of Information Security
Engineering
Soonchunhyang University
Asan-si, Chúngchóng-namdo,
Korea (Republic of)

Kathiravan Srinivasan
School of Information Technology
and Engineering
Vellore Institute of Technology (VIT)
Vellore, TN, India

ISBN 978-3-030-03507-5 ISBN 978-3-030-03508-2 (eBook)
https://doi.org/10.1007/978-3-030-03508-2

Library of Congress Control Number: 2018967450

This Springer imprint is published by the registered company Springer Nature Switzerland AG.
The registered company address is: Gewerbestrasse 11, 6330 Cham, Switzerland

To Amara Dewndarage (mum) and
Linton Jayakody (dad)
– Dushantha Nalin K. Jayakody

Foreword

The 5GPPP mobile networks incline to exploit the high-frequency technology to fortify a sizably voluminous number of contrivances at a higher rate without disruptions. With applications covering a massively enormous number of BSNs and IoT devices, a significant challenge of enhancing the security at all the layers of the network becomes an important objective. With a sizably voluminous number of contrivances operating at the same instance, it becomes arduous to identify security threats and intruders that can intentionally obnubilate in the prodigious information floating across the network. Security needs to be invigorated as more spectrum space is available on the 5G networks that can sustain heftily ponderous bits without affecting the transmission range. However, the enhancement of security requires a state-of-the-art set of protocols considering the architecture and operational overviews of the 5G networks. Furthermore, 5G communications are vulnerable susceptible to be hacked, and existing research results have alerted about the possibility of several types of attacks leading to a significant impact on the performance of these networks. The presence of infected codes, viruses, Trojans, wormhole, and malware in 5G devices can affect the entire network because these systems must operate instantaneously.

Academics, researchers, industrial partners, and other interested parties within local communities have developed and are implementing innovative approaches to enhancing the security of upcoming 5G networks. Further, some projects are happening around the globe which primarily intend to exploit the existing solutions for providing remedies to new challenges in 5G. However, the properties and characteristics of 5G networks make it considerably difficult to apply the existing technologies directly and demand new and competitive solutions which not only adhere to the demand but also perform well under different scenarios.

It is required that continuous efforts must be made to further reduce the obligations and performance issues cognate to the security of applications and contrivances in 5G networks. Subsisting mechanisms and solutions need to be ameliorated through more paramount solutions. Developing adscititious innovative approaches that can be applied to a broad set of devices and can facilely be utilized for further elongating the reachability of 5G technology should be the primary objective.

This book assembles the expertise of researchers and developers to present the state of the art in recent advances dealing with the security and privacy facets of 5G networks. The book comprises several chapters related to the core ideology of security, privacy, and trust in 5G-enabled wireless networks. The chapters cover diversified domains of security ranging from data security and privacy with physical-layer security. Also, there are studies related to the performance impact of security on the energy efficiency, which is one of the promising areas to be focused on futuristic security solutions. MIMO-based physical-layer security and application perspective are well-investigated in this book. Specific studies which highlight new paradigms, like catalytic and osmotic security, are also presented as a part of this book. The writing of the chapters and their unique compilation allow active researchers to have a detailed view of security perspective in 5G networks.

Especially from 5G's security point of view, it is mandatory to characterize the network requirements as well as understand the network layouts and factors affecting its performance. Regarding security, various models, protocols, adversarial attacks, key generation, and agreement in 5G applications and architectures must be studied. With the involvement of machine learning and artificial intelligence, the applicability of existing studies raises concerns, and it is desired to study them in a different angle while focusing the requirements of fast processing and reduced cost of ownership.

The editors of this book, Prof. Nalin D. K. Jayakody, Prof. Kathiravan Srinivasan, and Dr. Vishal Sharma, are well-known researchers in their community, and I know them as a part of my professional circle. Prof. Jayakody leads a lab as an Associate Professor at the School of Computer Science and Robotics, National Research Tomsk Polytechnic University, Russia. Dr. Srinivasan is an Associate Professor at the School of Information Technology and Engineering, Vellore Institute of Technology, Vellore, India. Dr. Sharma is a Research Assistant Professor in the Department of Information Security Engineering, Soonchunhyang University, South Korea. Their work in successfully compiling the chapters is commendable, and this book presents solid guidance for the community of both advanced and young professionals to follow ongoing and upcoming aspects of research results on security in 5G-enabled wireless networks. Thus, I recommend this book with an up-to-date and comprehensive text to the academic and industry and other interested parties.

With this, I wish them all the very best and congratulate them on the release of their book.

IEEE Fellow, ACM Fellow, AAAS Fellow, IET Fellow Yi-Bing Lin
Chair Professor, CSIE
National Chiao Tung University, Taiwan
September 18, 2018

Preface

This book emphasizes the security aspects of 5G networks. Traditional cryptographic security can only sustain until the computational power is not available to break the keys. Thus, it becomes of utmost importance to improve the architectures, which include placement of authentication server and the network core to protect data and connections from intruders and security threats. This book covers all the issues related to 5G network security.

The book provides details on network architecture and critical requirements. It also provides the issues concerning security overview, 5G architecture and policies, and various solutions which can handle these policies from the network-layer and physical-layer point of view. The book includes core principles of physical-layer 5G wireless communications solutions such as massive MIMO, energy harvesting, large multi-antenna systems and the use of millimeter wave (mmW) spectrum to form heterogeneous networks (HetNets), and so on to improve the performance of the 5G security protocols and models. Optimization of security models is covered as a separate section with a detailed section on the security of 5G-based edge, fog, and osmotic computing.

The chapter "5G Security: Concepts and Challenges" provides an overview of the issues and challenges in the security of 5G mobile communication. The security lies at the core of wireless communication. The issues become more intricate as 5G is aimed at diverse services ranging from high data rate to extensive connectivity. After an ephemeral portrayal of thc evolution of mobile communication, we embark on the discussion of security characteristics unique to 5G. Further, this is followed by the physical-layer planning of 5G and salient features of 5G, notably MIMO antenna arrays and cognitive radio. Security models are then presented. Topics covered include UE and radio network security, flexibility and scalability of security, network slicing security, and cloud security. Security protocols are an integral part of security. This chapter focuses on informal protocols. Several relevant concepts such as protocol authentication properties and message transmission channels are introduced.

A modest instance is used to illustrate the operation of the security protocol. Channel coding has been recognized as an effective approach to physical-layer security. In the last part of the chapter, several channel coding schemes that have been proposed for use in 5G are discussed. These include the well-known LDPC and polar codes. Finally, physical-layer security with passive and active eavesdroppers for several innovative techniques in 5G is addressed.

Chapter "5G Applications and Architectures" considers the present research direction in the fifth-generation (5G) network; it can merely draw the observation about the requirements of effective protocols, algorithm, and framework for steamroll transition. The chapter objective is to draw canvass as the futuristic adoption of application eagerly waiting for high data rate to carry those objectives with the development of protocols or openwork to resolute the use as well as security threats. In this chapter, the focus is on the application as well as the techniques which are likely to adapt to serve 5G. It begins with the brief introduction with the proper applications and promising methods. We have covered much promising technologies such as NOMA, massive MIMO, and millimeter wave with the architectural overview such as cross-layer.

Moreover, the importance and overview of SDN-NFV-based models have been explained which is helping massively in the changeover of transition. The significance of the work also enhances due to the research gap in existing proposal such as existing empirical model presented for 5G. Apart from the specific issues with the existing proposal, some general parameters (time, space, and hardware resources) with consequences are also listed. The limitation posed by the exciting technology has been given individually in the chapter where the proposal description has been made. This chapter provides insights into the cellular technology evolution and can be used as a guideline for technology development toward the fifth generation (5G). Finally, we conclude our chapter with good insight about industry initiatives as well as academic collaboration.

In chapter "A Survey on the Security and the Evolution of Osmotic and Catalytic Computing for 5G Networks," a comprehensive description of the security for the 5G networks is presented along with the discussions on the evolution of osmotic and catalytic computing-based security modules. The taxonomy on the basis of security requirements is presented, which also includes the comparison of the existing state-of-the-art solutions for the security of 5G networks. This chapter also provides a conceptualized security model, "CATMOSIS," which combines the features of catalytic as well as osmotic computing for enhancing the security of 5G networks. Furthermore, this idealizes the amalgamation of security features on the basis of catalytic and osmotic computing in the 5G networks. Finally, various security challenges and open issues are discussed to emphasize the works to follow in this direction of research.

In chapter "Physical Layer Security in 5G Hybrid Heterogeneous Networks", we discuss and analyze the physical-layer security perspective and performance of a massive multiple-input multiple-output (MIMO)-enabled hybrid HetNet with mmW small cells. Investigation of physical-layer security in such a network is an important topic since this network structure presents the most common deployment scenario for the 5G communication networks. We provide important insights about the secrecy outage probability, secrecy rates, secrecy energy efficiency, and secrecy spectrum efficiency at the end user, in the presence of eavesdroppers around, in the downlink communication scenario. Analytical results show that a higher density of small cells in the network leads to improved secrecy at the end user. Moreover, the number of antennas at the massive MIMO-enabled macro-cell BSs and beamforming directivity gains at mmW cells shall be carefully chosen for optimal secrecy performance of the network.

Chapter "Physical Layer Security of Energy Harvesting M2M Communication System" provides a broad introduction to M2M communications, discusses its applications, and highlights its critical challenges. Also, this chapter provides a discussion of the energy harvesting techniques and their importance in modern wireless communication systems. A detailed discussion of recent trends in physical-layer security (PLS) is presented. Using this basic knowledge of PLS and RF energy harvesting, we analyze the secrecy performance of an M2M communication system under the imperfect knowledge of wireless channel state. Later, the chapter provides a comparison of two jamming techniques, namely, full-duplex (FD) jamming and dedicated gaming. It is shown that the dedicated jamming performs better than the FD jamming at the cost of introducing an additional node with additional power requirements.

In the final chapter, the authors investigate the feasibility of co-time co-frequency uplink and downlink (CCUD) transmission in the cellular system with massive MIMO BS. Since massive MIMO has a high spatial resolution, it is not easy to eavesdrop and can improve the security of data transmission. Reliable beamforming is a prerequisite for high spatial resolution of massive MIMO. Therefore, this chapter focuses on how to perform reliability beamforming in CCUD systems. By exploiting the beam-domain representation of channels based on the basis expansion model, we prove that massive MIMO channel matrix (vector) can be represented by a low-dimension effective beam-domain channel matrix (vector). Based on this property, the authors propose a beam-domain full-duplex (BDFD) massive MIMO scheme (BDFD scheme for short) to enable a CCUD transmission in the cellular system. The authors show that the BDFD scheme achieves significant savings in uplink/downlink training and achieves the uplink and downlink sum capacities simultaneously as the number of BS antennas approaches to infinity. The authors, then, investigate several essential components for the practical implementation of BDFD scheme in the cellular system, including UE grouping, effective beam-domain channel estimation, beam-domain data transmission, and interference control between uplink and downlink. Finally, the authors examine the SE of BDFD scheme using the third-generation partnership project long-term evolution (3GPP LTE) simulation model of the macro-cell environment. The results demonstrate the superiority of BDFD scheme over the TDD/FDD massive MIMO.

Tomsk, The Tomsk Area, Russia — Dushantha Nalin K. Jayakody
Vellore, Tamil Nadu, India — Kathiravan Srinivasan
Asan, South Chungcheong Province, South Korea — Vishal Sharma

Acknowledgment

The growth of the telecommunication sector has improved the user experience regarding services and applications. With the upcoming 5GPPP, the mobile networks tend to exploit the high-frequency technology to support a large number of devices at a higher rate without disruptions. With applications covering a large number of BSNs and IoT devices, a significant challenge of enhancing the security at all the layers of the network becomes an important objective. With a large number of devices operating at the same instance, it becomes difficult to identify security threats and intruders that can intentionally hide in the vast information floating across the network.

Security needs to be strengthened as more spectrum space is available on the 5G networks that can sustain large bits without affecting the transmission range. However, the enhancement of security requires a new set of protocols considering the architecture and operational overviews of the 5G networks. With software-defined technology dominating the current era of telecommunications, it becomes essential to consider these for intelligent and user-centric security enhancement. Reduction of latency associated with the authentication of the users to the core as well as the authentication servers needs to be addressed for enhanced connectivity. Security should not come at the cost of coverage and capacity. The protocols and architectures which aim at providing security to users must not compromise with the factors affecting the capacity as well as the coverage. The use of novel devices for maintaining backhaul/fronthaul while comprehensively sustaining the security requirements of the networks is the need of the hour.

The editors would like to express their gratitude to all the contributors for their full cooperation during the entire authoring and production process and their endurance through the reviewing rounds and specifically to Hwang-Cheng Wang, Poorna Pravallika Sriram, Hema Ganesh Jami, Kathiravan Srinivasan, Dinesh Kumar Sah, D Praveen Kumar, Chaya Shivalingagowda, P.V.Y. Jayasree, Vishal Sharma, Gaurav Choudhary, Anum Umer, Syed Ali Hassan, Furqan Jameel, Muhammad Awais Javed, Dushantha Nalin K. Jayakody, Kui Xu, Xiaochen Xia, Yurong Wang, Wei Xie, and Dongmei Zhang.

The editors would like to acknowledge the funding support, in part, by the framework of Competitiveness Enhancement Program of the National Research Tomsk Polytechnic University, Russia. Finally, the editors would like to thank Mary E. James, Zoe Kennedy, and Abhishek Ravi Shankar from Springer for their support in bringing this book to completion.

School of Computer Science and Robotic
National Research Tomsk Polytechnic
Univcrsity, Russia
School of Postgraduate Studies
Sri Lanka Technological University, Sri Lanka
Dushnatha Nalin K. Jayakody

School of Information Technology and Engineering
Vellore Institute of Technology (VIT), India
Kathiravan Srinivasan

Department of Information Security
Engineering, Soonchunhyang University, South Korea
Vishal Sharma

Contents

5G Security: Concepts and Challenges

Poorna Pravallika Sriram, Hwang-Cheng Wang, Hema Ganesh Jami, and Kathiravan Srinivasan

Abbreviations

AKA	Authentication and Key Agreement protocol
AMTS	Advanced mobile telephone system
AN	Artificial noise
API	Application programming interface
ASM	Antenna subset modulation
BC-CM	Broadcast channel with confidential message
BER	Bit error rate
BS	Base station
BYOD	Bring your own device
CR	Cognitive radio
DNS	Domain Name System
DoF	Degrees of Freedom
EAP	Edge Automation Platform
eMBB	Enhanced Mobile Broadband
EPS	Evolved Packet System
ETCI	European Telecommunications Standards Institute

P. P. Sriram (✉)
Department of Electronics & Communication Engineering, Vel Tech Rangarajan Dr. Sagunthala R&D Institute of Science and Technology, Chennai, India

H.-C. Wang
Department of Electronic Engineering, National Ilan University (NIU), Yilan City, Taiwan

H. G. Jami
National Ilan University, Yilan City, Taiwan

K. Srinivasan
School of Information Technology and Engineering, Vellore Institute of Technology (VIT), Vellore, TN, India

D. N. K. Jayakody et al. (eds.), *5G Enabled Secure Wireless Networks*,
https://doi.org/10.1007/978-3-030-03508-2_1

ETSI	European Telecommunications Standards Institute
FDD	Frequency Division Duplex
FDMA	Frequency Division Multiple Access
HD	High Definition
HIP	Host Identity Protocol
HTTP	Hypertext Transfer Protocol
IEC	International Electro-technical Commission
IEEE	Institute of Electrical and Electronics Engineers
IETF	Internet Engineering Task Force
IMSI	International Mobile Subscriber Identity
IMT	International Mobile Telecommunication
IMTS	Improved Mobile Telephone System
IP	Internet Protocol
IPv4	Internet Protocol version 4
IPv6	Internet Protocol version 6
IPWAVE	Internet Protocol Wireless Access in Vehicular Environment
ISO	International Organization for Standardization
ITU	International Telecommunication Union
KKT	Karush-Kuhn-Tucker
LDPC	Low-density parity check
LTE	Long-Term Evolution
M2M	Machine to machine
MA-WC	Multiple access wiretap channel
MIMO	Multiple input, multiple output
mmWave	Millimeter Wave
MPWG	Mobile Platform Work Group
MTS	Mobile telephone system
NERC	North American Electric Reliability Corporation
NOMA	Non-orthogonal multiple access
NP	Network planning
ONF	Open Networking Foundation
PTT	Push to talk
QoS	Quality of service
RAN	Radio access network
RS	Relay stations
SDO	Standards Development Organization
SDR	Software-defined radio
SIC	Self-interference cancellation
SINR	Signal-to-interference-noise ratio
TCG	Trusted Computing Group
TCP	Transmission Control Protocol
TDD	Time division duplex
UAV	Unmanned aerial vehicle
USIM	Universal Subscriber Identity Module
WG	Working group
ZFBF	Zero-forcing beam-forming

1 Overview

1.1 Introduction

Wireless mobile communication has initiated its technology formulation, revolution, and development from the 1980s. In the last few decades, wireless mobile innovations have experienced third and fourth generations of technology development and revolution. The quality of services (QoS) and security are significantly promoted in 4G, while the cost per bit is low. In comparison with the previous network generations, there are some issues with 4G such as greater power consumption (battery use) and the high cost of the equipment needed to implement the next-generation network. 5G is going to be the next generation. Mainly, it aims to provide a complete wireless communication with almost no limitations. As six billion people own smartphones, we are going to dissect the different generations of cellular technologies. Moreover, the digital wireless communication systems are relentlessly determined to satisfy the developing need for individuals. Also, in the 5G technology, the rate of the data calls is made easy compared to the previous generations as the quality of the service is excellent and highly flexible and has significant spectrum management and improved efficiency with decreasing cost. This whimsical development represents not just the powerful need for individuals all over the globe to communicate and connect with each other and also to have access to the information but also the gigantic steps that innovation has made in satisfying the need [1]. All IP-based fourth-generation Long-Term Evolution (LTE) networks have become a portion of the day-to-day routine with the rapid rise in demand of the smart mobiles. As a result, a set of new user-oriented mobile multimedia applications, like video conferencing, streaming video, e-health care, and online gaming, has arisen. Furthermore, as the sphere prepares for the first commercial debut of 5G networks, many people are inquisitive about the security hazards and risks. 5G technology will reinforce a massive number of connected devices, which enables a colossal rise in bandwidth and create a next-generation hazard landscape that will irresistibly introduce 5G security challenges. These new applications are fulfilling user necessities as well as opening the new business skylines for remote operators to expand their income.

1.2 Evolution of Cellular Technologies

The evolution of the mobiles classifies various technologies into different "generations." The word "generation" indicates the change in the quality of service, adaptable transmission technology, and new frequency bands.

First Generation

The acronym 1G represented the first-generation mobile telecommunication and was first introduced in the 1980s and continued till 1990. These networks use analog systems for communication. Mobile devices are simple voice-only cellular phones. Moreover, the first generation of analog mobile phones has a speed of 2.4 Kbps and 14.4 Kbps. It allows its consumers to make voice calls only within the same country. Voice call modulation is performed using a technique called frequency division multiple access (FDMA). It proposes many mobile technologies like mobile telephone system (MTS), advanced mobile telephone system (AMTS), improved mobile telephone system (IMTS), and push to talk (PTT) [2]. On the other hand, it has low capacity, deceptive hand-off, and weak voice links and does not offer security.

Advantages of 1G technology
1G represents the initial success in the attempt to achieve mobile communication. In retrospect, 1G has few advantages. Nevertheless, it signifies an important step in the development of mobile technology, and many of the fundamental ideas such as cells, frequency reuse, and multiple access have remained in subsequent generations of mobile communication.

Limitations of 1G Technology
- Constrained limit: It has a very little coverage area cellular network.
- Low calling limit: The quality of the call service is low because of its low capacity.
- No space for range development: There is no room for the spectrum growth.
- Poor data communications: The data rate speed is low.
- Negligible security: Privacy and security assurance is minimum.
- Deficient extortion security: Protection toward fraud sites is not guaranteed.

Second Generation

2G refers to the second generation, which is based on the global system for mobile communication (GSM). 2G networks use digital signals, and its data speed is in between 14.4 Kbps and 64 Kbps [3]. This network offers unique services such as short message services (SMS), picture messages, and multimedia messages (MMS). It cannot handle complex data such as video which is the most notable drawback. The network capacity of 2G is very much better than 1G. The primary distinction between 1G and 2G is that 1G uses analog signals whereas 2G uses digital signals for communication.

Advantages of 2G Technology
- Improved privacy is the added advantage of 2G technology.
- 2G technology introduces the digital data services such as SMS and email that have allowed the world to shrink and people to get closer.

- It enables users to place a call on hold in order to access another call.
- The digital data service is used to assist the mobile network operators to introduce short message service over the cellular phones.

Disadvantages of 2G Technology

- 2G technology requires powerful digital signals to help the mobile phones work, but the digital signals could be weak if there is no network coverage in any specific area.
- The data rate is low. The downloading and uploading speeds available in 2G technologies are up to 236 Kbps.

Limitations of 2G Technology

- It demands intense digital signals to assist the connections of mobile phones.
- Complex data types such as videos are not supported.

Third Generation

In the third generation, the wireless communication terrace has voice and data potency. It was established in 2000. 3G is the first international standard system released from ITU, in divergence to the previous generation systems. It works in frequency division duplex (FDD) and time division duplex (TDD) modes. As compared to 1G and 2G, it provides a higher speed which ranges from 144 Kbps to 2 Mbps. It has a bandwidth of 25 MHz. Also, it is referred to as Mobile Telecommunication 2000. It introduced data services, expanding the functionality beyond voice and including multimedia, text, and some limited Internet access. The foremost technological dissimilarity that differentiates 3G technology from 2G technology is the use of packet switching rather than circuit switching for data transmission [4].

Advantages of 3G Technology

The 3G network uses a wide range of radio spectrum that allows faster data transmission. It also allows location-based services like weather reports on the mobile phones.

Limitations of 3G Technology

- The price for 3G services is expensive – It provides better-quality services compared to 2G technology services.
- Expensive in nature – Due to the voice and data rate services, the rate of 3G is a little more expensive.
- Higher bandwidth requirements – The requirement for the bandwidth is high due to the heavier usage of data calls.

Fourth Generation

The fourth generation of mobile communication is a packet switched wireless system with colossal area coverage and high throughput. It is designed to provide high spectral profitability. It provides communication with higher data rates and high-quality video streaming in which Wi-Fi and WiMAX are combined [3]. This network can provide a speed of upto 50 Mbps. The quality of service (QoS) and security are symbolically promoted in 4G, while the cost per bit is low.

The primary 4G protocol – LTE – was designed to reinforce the mobile broadband and is the dominant industry standard today. The significant frequency band range is between 2 and 8 GHz. It also provides the ability for worldwide roaming to allow access to cellular communication anywhere.

Advantages of 4G Technology
The most apparent benefit of the 4G mobile network is its prodigious speed. Expanded bandwidth leads to much higher data transfer speed, which is particularly advantageous for mobile gadgets. 4G offers coverage of 30 miles and more. The uploading speed in 4G is upto 5 Mbps and the downloading speed is upto 50 Mbps.

Limitations of 4G Technology
- It is expensive and hard to implement.
- It demands more battery usage.
- It needs complex hardware.

Fifth Generation

This 5G technology merges all the upgraded benefits of mobile phones like high-speed dialing, music recording, cloud data storage, and high-definition (HD) downloading instantaneously. New radio bands above 20 GHz are being designated for 5G. 5G networks are also intended to meet new use cases, such as the Internet of Things, services, and lifeline communication in times of natural catastrophe. It will be crafted for an extraordinary system to broadcast immense amount of information in gigabits per second (Gbps), enabling media news feeds and TV programs with HD quality [3].

Even though 4G has not been around for a very long time, it is found to be inadequate in dealing with the various necessities in terms of denser networks and increased capacity factors such as the widespread use of smartphones, in terms of data rates, speed, coverage, battery life, and the emergence of the Internet of Things (IoT). This is not the technology's flaw; the smartphone revolution had not started when the 4G requirements and technologies were considered and selected. New applications are always developing. Nevertheless, overcoming the current limitations of 4G is the primary goal of 5G. The concept is to meet future demands for data rates, speed, coverage, and battery life in architecture to enable a cost-effective network that can be efficiently scaled.

Issues and Challenges of 5G

The key challenges in meeting the performance of these future networks using affordable technologies that are still in research and investigation are:

- Number of supportive devices
- Data volumes
- Lower cost with higher capacity
- New air interface
- Speed

Features of 5G Technology

- Faster data transfer rates compared to the previous generations
- Large memory
- Swift dialing speed
- HD quality impression
- More attractive and effective
- Peak uploading and downloading speed
- Remote diagnostics
- Up to 25 Mbps connection speed
- High-quality services to prevent errors
- Bidirectional large bandwidth

As the world embraces for the first commercial debut of 5G networks, many people are contemplating about the security hazards and risks that the new standard is going to encounter. 5G networks will feature a vast number of connected devices, see a substantial increase in bandwidth, and create a next-generation hazard landscape that will inevitably introduce unique security challenges [5].

1.3 The Significance of 5G Security

The 5G standard will bring the towering benefits such as upgraded speed and performance, low latency, longer battery life, greater capacity, and superior efficiency. Further, the 5G networks are not only preferred for faster data rates but also provide a backbone for many new services in the networked society, such as IoT and the industrial Internet. These services will provide connectivity for autonomous cars and unmanned aerial vehicles (UAVs), remote health monitoring through body-attached sensors, smart logistics through item tracking, remote diagnostics, and preventive maintenance of equipment [6]. However, immensely increased number of devices and high usage of virtualization and cloud will lead to many multifaceted 5G security threats, hazards, and attacks. Moreover, to realize healthy and robust communication in the future, the industry should strive to maintain a high standard of 5G security.

1.4 The Need for Security

The strategies to attack telecommunication systems have changed from war dialers to viruses to modern-day sophisticated, steady risks. The devices to protect multimedia transmission have likewise advanced from physical access control to modern applications and firewalls. Expanded utilization of cell phones for data services and applications has presented these gadgets with similar security dangers that were once known and confined to personal computers (PCs). Bring your own device (BYOD) and cloud technologies have additionally lifted the enterprise limits and regularly invited security specialists to work out novel solutions [7]. Mobile devices have not yet replaced personal computers but have turned into a perfect place where individual data can be found for depraved use. Therefore, security should be architected to shield from the present dangers as well as to address the expanding and developing risk landscape. Sufficient security ought to incorporate danger intelligence, visibility, and real-time protection [7].

Likewise, seeing to the presumable results of an assault, the harm may not be restricted to a business or notoriety, it could even severely affect open security. Besides, this prompts a need to increase particular security functional zones. Attack resistance should be a plan thought when characterizing new 5G protocols [8]. Applications for 5G are beyond traditional mobile connectivity needs to new public communication, IoT, smart world based on smart cities, smart transportation, and more. One of the principal challenges for 5G adoption is security-related challenges. Although security is a mandatory requirement of 5G networks, many of the 5G security-related issues arc still under development. However, the rapid adoption of 5G network will soon raise the requirement of a comprehensive handbook of 5G security.

2 5G Security Standardization

In February 2017, 3GPP published "Service Requirements for the 5G System" (TS22.261) that defines performance targets in various scenarios such as indoor, urban, rural, and different applications (intelligent transport, remote monitoring, and so on) [9]. 3GPP publishes 5G Phase-1specifications in 2018 as Release 15 and plans to publish Phase-2 in 2020 as Release 16. Since 5G is expected to be completely converged with Internet protocols, the standards produced by the Internet Engineering Task Force (IETF) are expected to play a vital role [9].

2.1 Internet Engineering Task Force

The IETF is an open standards organization that develops and promotes the center Layer-3 and higher protocols for the Internal and Intranet standards. It has existed as a formal Standards Development Organization (SDO) since mid-1986, yet the principal guidelines and demand for remarks had been formulated in the late 1960s. A portion of the eminent IETF benchmarks include the Internet Protocol form IPv4 and IPv6, Domain Name System (DNS), Transmission Control Protocol (TCP), and Hypertext Transfer Protocol (HTTP) [8]. The 3GPP has incorporated a large number of IETF protocols in the cellular framework design throughout the years. IETF and 3GPP keep in close contact to track specialized topics.

The relevant working groups are, for example, IP Wireless Access in Vehicular Environments (IPWAVE) working group (WG) and Host Identity Protocol (HIP) working group on secure mobility protocols. If 5G networks serve safety crucial applications as envisaged, the ISO (International Organization for Standardization) will introduce standards such as Common Criteria (ISO 15408). For instance, for car connectivity, a specific standard is ISO 26262, which covers car safety requirements. ETSI (European Telecommunications Standards Institute) was the initiator of GSM standard and critical contributor of W-CDMA as 3G standard within 3GPP. As a co-founder of 3GPP, ETSI is actively involved in developing 5G through organizing such events as ETSI Summit on 5G network infrastructure, which focused on 5G standardization in 2017 [10]. ETSI identified priority applications for 5G as mobile broadband evolution, massive M2M communication, and ultra-reliable low-latency communication.

ETSI is also a known contributor to the Network Functions Virtualization Industry Specification Group (ISG) and is currently forming a group focusing on 5G security [9]. ITU (International Telecommunication Union) receives input from regional organizations such as ETSI in Europe and ARIB in Japan and develops recommendations for standards defining bodies. ITU Telecommunication Standardization Sector (ITU-T) created a Focus Group on International Mobile Telecommunications (IMT-2020), which operated in 2015–2016 and analyzed requirements and framework for the 5G ecosystem [10]. ITU Study Group 17 (SG 17) focuses entirely on security aspects of telecommunication. Several other relevant standardization bodies include IEEE 802, TCG, and ONF [10]. Interoperability and mobility with third-party networks such as Wi-Fi involve standards from the IEEE (Institute of Electrical and Electronics Engineers) such as 802.11. At Trusted Computing Group (TCG), the Mobile Platform Work Group (MPWG) develops use cases, frameworks, and analysis of 5G security. Open Networking Foundation (ONF) promotes the use of software-defined networking protocols and network operating systems. Its specifications, including OpenFlow, could become part of 5G core architecture and therefore are also important from the security point of view.

3 Security Characteristics of 5G

Basic Security Characteristics of 5G

- *Prevention of threats*: Reducing the ground issues for most security incidents. The firewalls are used to protect the network and control access to reduce the user-based risk. Intrusion detection and prevention tools are considered for blocking basic 5G security threats.
- *Terminating and fixation of advanced malware*: Going beyond signature-based tools helps to spot the attacks designed to evade basic filters. Behavior-based checks on endpoints – possibly using sandboxing – are important. Once a threat is detected, all the instances of it on the network should be removed.
- *Detecting anomalies:* Usage of the packet capture, big data, and machine learning to identify threats that are not detected by the basic filters. When inserted into the network switches and routers, it is more effective as it turns those devices into 5G security sensors.
- *Incorporate DNS (Domain Name System) intelligence*: DNS activity is monitored and protected against any malicious attacks.
- *Making threat intelligence paramount*: In order to understand the malicious efforts of hackers, providers must look for vendors that can profile hackers.

3.1 Drivers of 5G

The drivers for security have set up to give a reliable fundamental availability benefit. This fundamental trust will keep on being a driving force for 5G that organizes, as a high information rate, portable broadband administration. 5G systems will not be composed exclusively to provide new capacities for individuals and society but also to interface ventures (e.g., assembling and preparing, intelligent transport, and smart grid). With 5G, it is possible to foresee new models of how network and transmission services are provided. For instance, an automaker may wish to give management services to vehicles. Setting up coordinate wandering concurrences with different access arrangement for suppliers could be a cost-effective approach to accomplish this. Correspondingly, the idea of terminal/gadget will change: unattended machines and sensors will associate. Moreover, at times whole hairlike systems containing one or many individual gadgets will connect to the 5G network. Cloud and virtualization advancements will be utilized to lower the costs and improve the benefits more quickly. Telecom systems will uncover application programming interfaces (APIs) [8] toward users and outsider specialist organizations to a higher degree. Besides, general consciousness of user security in the public eye has expanded, prompting a more prominent spotlight on the assurance of user meta-data and correspondence. This issue turns out to be significantly more focal with the improvements in big data analytics. What describes 5G, considerably more than 4G, is that it will have a significant part in the task of society. The full extent of security, protection, and flexibility will be a change that comes a long way past innovation. It will impact legal systems and direction and activities of business elements and people.

3.2 Significance of Security and Privacy

The four characteristics of 5G networks and their usage, each with suggestions for security and protection, are:

- Modern confide models
- New relevance transmission models
- Emerging risk prospects
- Raised privacy concerns

Modern Confide Models

Confide models change over time. For instance, consider bring-your-own-device (BYOD) tendency in business. Beforehand, all user gadgets were thought to be reliable, as they were the greater part of a similar kind, all issued and overseen by the corporate IT division. Today, users need to utilize their own gadgets, thereby causing dangers like potential Trojan horses behind corporate firewalls [8].

Since 5G is going to support new plans of action, trust models will change. For example, for new basic administrations, what security necessities will be anticipated onto the 5G systems? The new kinds of gadgets will traverse a greatly extensive variety of security necessities and will in the meantime have completely different security requirements. Gadgets have so far been accepted to consent to models and not to intentionally endeavor to assault networks.

The current confide model does not explicitly capture this evolving business and technological scenery of 5G. To guarantee that 5G can bolster the requirements of new plans of action and guarantee adequate security, the trusted display outline is redrawn [8]. In that capacity, this does not really mean totally upgrading security. Nonetheless, it is urgent to distinguish any huge weaknesses.

Security for New Relevance Transmission Models

The utilization of clouds and virtualization underlines the reliance on protected programming and prompts different impacts on security. Current 3GPP-characterized frameworks depend on practical hub details and unique interfaces (reference focuses) among them, and all things considered, give a decent beginning stage to virtualization. Recently, committed/exclusive equipment has still regularly been utilized for these hubs and interfaces. Decoupling hardware and software implies that telecom software can never again depend on the particular security properties of a committed telecom hardware. For a similar reason, standard interfaces to the computing/network stages – for example, those characterized by ETSI (the European Telecommunications Standards Institute) in their Network Functions Virtualization work – are important to guarantee a sensible way to deal with security [8]. At the point where operators having third-party applications in their communication fabric and executing on hardware indistinguishable from local telecom services, there are demands on virtualization with solid detachment properties.

Emerging Risk Prospects

5G networks will considerably play an important part as the basic foundation. Numerous individuals have experienced events when phone lines, web access, and TV have all stopped working at the same time during a large-scale system blackout. The quality facilitated in and created by the 5G framework is evaluated to be much higher, and the advantages (hardware, software, and information) will be considerably more appealing for various sorts of attacks. Besides, thinking about the conceivable outcomes of an assault, the harm may not be restricted to a business or community; it could even severely affect the society as a whole. This prompts a need to reinforce certain security measures. Attack resistance should be a key ingredient when designing new 5G protocols [8]. Faulty validation strategies, for example, username/password should be eliminated. However, in a general sense, the new risks stress the requirement for quantifiable security affirmation and consistency; in other aspects, checking the presence, accuracy, and adequacy of security capacities is also important. Those utilizing 5G will require answers to inquiries, for example, is it safe to set up a virtual machine on a given equipment? What security tests have been conducted on a product? A key resource of the Networked Society will be information. As the carriers of information, 5G networks should give satisfactory assurance as separation and productive transport of ensured (scrambled/verified) information. The all of 5G gadgets and network won't simply influence innovative attack patterns; the social engineering attacks will likewise increase. Individuals asserting to be work associates or repair experts, for example, may contact an individual and demand different sorts of access to the person's data as well as to his/her gadgets.

Raised Privacy Concerns

There have been a few recent news stories that reported fraud base stations tracking users in urban communities and extracting individual information without user's knowledge [8]. The security of individual information has been examined inside EU. It is being audited in standardization bodies, for example, the 3GPP and the

IETF, and debated in many other forums. An especially delicate resource is the user identifier(s). As far back as 2G, user security has been a huge concern. However, the International Mobile Subscriber Identity (IMSI) assurance has so far only provided limited protection.

4 Network Planning

The steady growth in demand for better mobile experience, higher data rates, and lower latency is promoting the development of the upcoming generation of wireless systems, namely, 5G [11]. Network planning (NP) is one of the essential stages in deploying a wireless network that meets certain coverage, capacity, and quality of service (QoS) requirements. The planning process can minimize and optimize the locations of base stations (BSs) in a selected geographical area [11].

The precise planning for network establishment consists of:

- Preplanning
- Detailed planning
- Post-planning or optimization

Preplanning The output of preplanning is a surmised number of base stations required to cover an area of interest.

Detailed Planning The detailed planning stage permits the decision of the actual positions of the BSs inside the zone to be served [12].

Post-planning In the optimization stage, which happens after the system has been deployed and is running, the network performance is inspected, potential problems are detected, and network operations are enhanced by the improvement [12].

4.1 Objectives

The objective of NP mostly depends on the business strategy of the operators. The coverage targets for different types of services, taking into account billing and throughput policies, regulatory constraints, market share goals, and competition [11]. Ultimately, the objectives can be boiled down to the following set of optimization goals identified in the cell planning phase.

1. *Minimize TCO*: In addition to minimizing the overall network cost, this objective may also include minimizing economic costs related to deployment and parameter optimization.

2. *Maximize capacity*: For a single service, this goal can be defined as the number of clients who can be served at once. On account of multi-service traffic, the capacity can be approximated with respect to worldwide throughput.
3. *Maximize coverage*: This includes satisfying coverage policy requirements for various services. Uplink (UL) and downlink (DL) coverage must be balanced. Both traffic channels and coverage of standard channels must be considered [13].
4. *Minimize power consumption*: Health concerns have motivated the radiated power minimization objective. However, the recent awakening of a desire for greener wireless systems has added more depth to this objective. Consequently, power consumption, including fixed circuit power as well as variable transmission power, must be minimized.

Optimize Handover (HO) Zones In a well-planned cellular system, a certain proportion of the area of each cell should overlap with neighboring cells to satisfy HO conditions. HO zones are essential to ensure the continuation of service between sectors. It also strengthens the radio connection against fast fading and shadowing [14]. However, too much overlap may result in wastage of power and radio resources and increase in interference and electro-smog, making it a tricky planning objective.

4.2 Planning Inputs

Various inputs are required to solve the cell planning difficulty depending on objectives in focus and phase of planning. The following inputs need to be known for planning the network.

Traffic Models

User traffic distribution is a primary factor that ultimately determines the cellular system plan and, hence, is a crucial input in the network planning process. In GSM (mono-service systems), for instance, geographical characterization of traffic distribution is sufficient. However, with multi-service systems supporting data, traffic characterization based on types and level of service is needed. Test point-based traffic models are often used for network planning traffic modeling, for the sake of practicality [15, 16]. In this model, an area is characterized by a time interval, and all located mobile terminals are bundled into a single test point [17]. This point represents the increasing traffic, or traffic intensity, from all those terminals, over the determined interval.

Potential Site Locations

Theoretically, a base station can be installed anywhere. However, in the real world, a set of candidates is first predetermined and used as input to the cell planning, to incorporate the real estate constraints [15]. The objective is to find the optimum subset of base station locations. These potential BS locations are determined by taking into account constraints such as socioeconomic feasibility and availability of site(s), traffic density, building heights, terrain height(s), and preexistence of a site(s) by the same or other operators.

BS Model

There are many parameters that define the base station model, like antenna type and height, receiver sensitivity, load capacity, transmit power, and capital and operational costs [16]. Moreover, heterogeneous networks necessitate modeling of new types of nodes, for instance, relay stations (RS), picocells, femtocells, and small cells.

Propagation Prediction Models

A key input to the planning process is the signal propagation model. The potential of this model is to incorporate reflection, diffraction, absorption, and propagation of the signal in a real environment [17]. The natural and human-made structures, vegetation, and topography of an area largely determine the accuracy of the network planning outcomes. Very sophisticated planning tools rely on actual measurement-based propagation maps, or ray tracing-based complex analysis, to predict the propagation. However, obtaining complete propagation maps of a large area using these methods is a very cumbersome, time-consuming, and expensive process [15]. For this reason, different empirical models have been proposed in the literature. Such models abstract the experimental and statistical data in the form of deterministic expressions that can easily be used in network planning. Okumura, Hata, and COST 231 are a few examples of such well-known propagation models used in network planning to depict propagation loss in different environments and scenarios [16]. Fine-tuning of these models is done by setting parameters within these models to reflect the real-world conditions as closely as possible. While propagation models for sub-5 GHz frequencies are well established, research on developing such models for higher frequencies such as mmWaves is still in progress [18].

4.3 *Planning Outputs*

The network planning process intends to provide one or more of the following outputs:

- The ideal number of base stations
- The best areas to fix base stations
- The kind of base station ideal for each area
- The configuration of parameters such as antenna height, number of sectors, and sector orientation, tilt, and power
- Frequency reuse pattern
- Capacity dimensioning, e.g., number of carriers or carrier components per sector

4.4 *Types of Network Planning*

The objectives, input, and output of the network planning process also depend on the type of planning. There are two types of network planning as described in the follows.

Rollout Network Planning

This is the network planning where no prior networks exist, and a logical state approach can be used to meet all the objectives of interest. Regarding input parameters, in this phase, the traffic distribution is not exactly known yet [18]. Estimates of traffic based on geo-marketing forecasts are used for planning in this phase.

Incremental Network Planning

This type of network planning is carried out after the first rollout planning to meet the increasing demand. Unlike the plane state approach, planning in this phase is bounded by additional constraints imposed by existing sites. However, in this phase, the traffic distribution can be modeled with much better accuracy using measurements from existing network reports. It is anticipated that 5G deployment will mostly require incremental planning by building on LTE/UMTS/GSM network [18].

5 5G Roadmap

5G is not just a progressive upgrade of the previous generation of cellular but a revolutionary technology envisioned to defeat the bounds of access, bandwidth, performance, and latency limitations on connectivity worldwide. It has the potential to enable fundamentally new applications, industries, and business models and dramatically improve quality of life around the world.

5.1 Need for Roadmap

Remarkably, it is fundamental to comprehend which innovative disruptions are required to empower mobiles to last as well as to flourish in an undeniably competitive technology and business landscape. Understanding that innovative disruption is firmly coupled to advancement, this community IEEE innovation roadmap lays out the technology and development vision for the telecommunications industry and a significantly more far-reaching industry stakeholder system. With appropriate direction, it is foreseen that 5G and beyond will have the capacity to realize the financial advantages envisioned in various studies [19].

The critical benefits of the IEEE 5G technology roadmap are to:

- Focus endeavors toward future solutions, with the goal that advantages are boosted for the industry.
- Amortize R&D costs through coordinated efforts and associations.
- Analyze unique developments to obtain potential solutions which serve stakeholders in the business.
- Align with peer-competitive arrangements that can be executed in synergistic environments and also in the competitive domain.
- Contribute and be educated on normal points of view to address the mutual needs and difficulties involved in the evolution to the future state.
- Empower visibility into future innovation patterns.
- Enhance venture methodologies for R&D.
- Make important contributions to the standards.

Building up a time allotment of projections when presenting new innovations might be needed to convey basic advantages like:

- Gives essential lead time to gear and interface improvement
- Enables time for solutions to be displayed and tested
- Empowers research opportunities to be explored and financed

5.2 *Roadmap Process*

In mid-2016 the community found that there is a need to build up an arranged vision for the wireless connectivity ecosystem. The IEEE 5G Initiative was hence forth presented under the sponsorship of IEEE with a commencement meeting in August 2016, in Princeton in the United States [19]. Also, a significant workshop was held in conjunction with IEEE GLOBECOM 2016, where the vision was enunciated in broad daylight and supporters were urged to contribute to the developing roadmap ecosystem. From these workshops and conferences, a working philosophy has been set up. Generally, the focus of this IEEE innovation guide is to recognize basic needs, difficulties, and potential solutions/zones of development. The point is to build partnerships and coordinated efforts among industry groups, be comprehensive of all divisions of the wireless community and to be driven by industry trends. The goal is to refresh it intermittently; in particular, an arrangement of advance reports is to be routinely delivered and conveyed to the business at large.

The following topics reflect the roadmap working group:

- Massive multiple input, multiple output (MIMO)
- Hardware
- Edge Automation Platform (EAP)
- Millimeter Wave (mmWave)
- Security
- Standardization building blocks
- Applications and services

6 Existing Concepts of 5G

6.1 *Multiple Input and Multiple Output*

Using multiple transmit and receive antennas, MIMO method is used to increase the capacity of the channel in radio link. MIMO has turned into an essential component of remote correspondence framework benchmarks including IEEE 802.11ac (Wi-Fi), WiMAX (4G), and so on [20]. The advantage of MIMO is that a greater amount of data can be sent across the wireless channels, thereby improving the energy efficiency, spectral efficiency, and reliability. In MIMO configuration, both the transmitter and receiver sides may contain a huge number of antennas [21]. In the current days, "MIMO" refers to a pragmatic system for sending and accepting in excess of one information stream on a radio channel simultaneously through multiple propagation paths, as illustrated in Fig. 1.

In a MIMO system, the transmitting antennas as well as the receiving antennas are distributed to many devices. Further, one of the main benefits of MIMO technology is that intracellular interference and noise can be reduced. Because of these benefits, MIMO is considered to be a key technology.

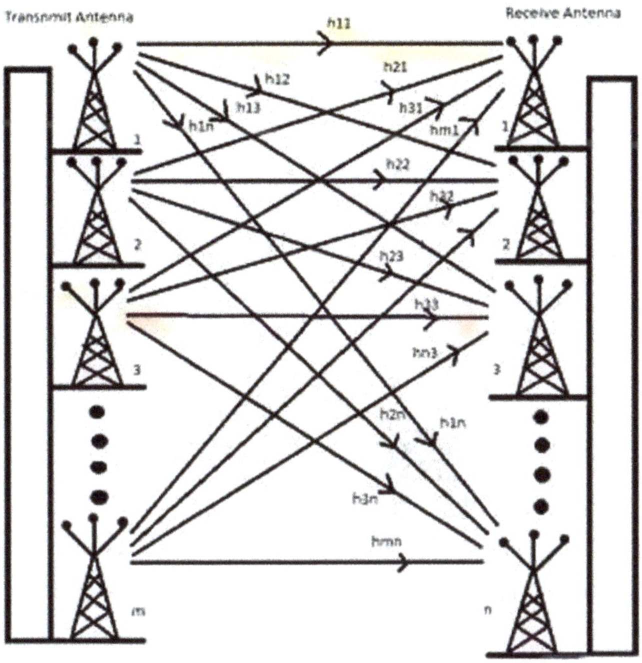

Fig. 1 MIMO system

6.2 Cognitive Radio Network

Cognitive radio (CR) is a vigorous, keen radio and system technology that can automatically recognize available channels in a remote range and furthermore change transmission parameters, thereby enabling more communications at the same time and enhancing radio-operating behavior. Innovations like adaptive radio and software-defined radio (SDR) [20] are utilized in a CR network. Adaptive radio is a technology where the communication system operates and changes its performance. As for SDR, various hardware components like modulator, demodulator, amplifier, and mixer are replaced by an intelligent software system [22]. Cognitive radio network is used to improve the utilization of radio-frequency spectrum as shown in Fig. 2.

As shown in Fig. 3, in one cognitive radio cycle, a device monitors the spectrum bands to detect blank spectrum spaces (white spaces or holes). From the qualities of the spectrum spaces that are recognized through spectrum sensing, a suitable range of band is picked based on the radio characteristics and user requirements [22]. If a band of the operating spectrum is determined to be available for use, communication can be executed over that spectrum band.

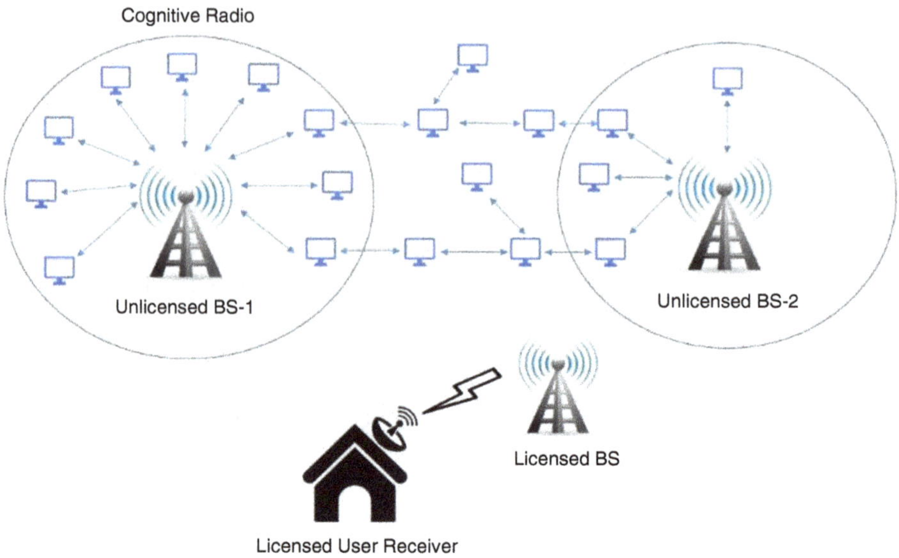

Fig. 2 Cognitive radio system

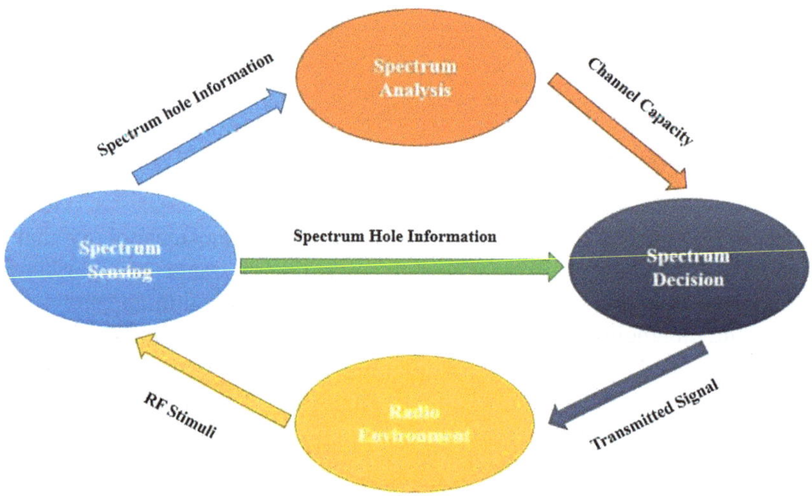

Fig. 3 Cognitive radio cycle

7 Security Models

5G frameworks are the next stage in the advancement of mobile communication and will have a crucial empowering influence on the Networked Society [8]. This advancement engenders new security situations and requires new security arrangements. 5G networks will bring a massive number of connected devices, interoperability of new and legacy access technologies, significant increase of bandwidth, and new business cases that will usher in new challenges from the security perspective [23]. 5G security will not only be defined by quantitative aspects such as bit-rates and latency but also by subjective perception, for example, new business and trust models, better approaches for delivering services, a developed risk landscape, and an expanded concern for protection.

7.1 *Identity Management*

The 4G LTE standard needs USIM (UMTS Subscriber Identity Module) on the universal integrated circuit card to gain network access [8]. This way of handling the identity will continue to be an essential part of 5G for reasons such as abnormal state of security and ease of use. Embedded SIM has likewise brought down the bar for organization issues identified with machine-to-machine communication. In any case, there is a general pattern of bring-your-own-personality, and the 5G ecosystem would for the most part benefit from a more open identity administration system. The risk of IMSI retrieval, where fraud radio system hardware demands cell phones to reveal their personality, has been talked about amid the 3G and 4G institutionalization process [23]. In any case, no insurance instrument was presented around them, as the anticipated dangers did not appear to justify the cost or multifaceted nature involved. It isn't clear whether this hazard examination is as yet legitimate and improved IMSI security merits thought for 5G.

7.2 *UE Security*

In the next-generation system, the storage of credentials and identities for both human and machine-type devices is required in the UE. The credentials and identities may be stolen from attacks by software or hardware. Such security threats can impact the subscriber or operator network [23]. 3GPP SA3 has currently agreed that a secure element for credential storage in UEs must provide:

1. Integrity protection of the subscription credential(s)
2. Confidentiality protection of the long-term key(s) of the subscription credential(s) (e.g., Key (K) in EPS AKA (authentication and key agreement)) [23]
3. Execution of the authentication algorithm(s) that make use of the subscription credentials

The above requirements should be achieved within the UE, with the use of a tamper-resistant secure hardware component. Implementations of these requirements shall allow security evaluation/assessment. The Subscriber Identity Module (SIM) functions for 5G, i.e., next-generation USIM will inherit from previous standards [23]. In a similar manner to LTE system, the next-generation USIM will be able to generate symmetric keys. It may also be able to generate new asymmetric key pairs and even new trusted public keys.

7.3 Radio Network Security

Due to the emerging risk landscape and new innovation that furnishes users with minimal effort to program their own gadgets (even at radio access level), the attack resistance of radio systems ought to be an all the more blunt plan thought in 5G, examining risks, for example, Denial of Service from conceivably getting out of hand gadgets, and adding moderation measures to radio protocol outline [23]. In spite of the fact that LTE radio access has phenomenal cryptographic assurance against eavesdropping, there is no security against changing or infusing user plane traffic. With 5G radio access as a building block in, for instance, mechanical automation, the potential advantages of including integrity assurance appear to be deserving of investigation [8].

7.4 Flexible and Scalable Security

With virtualization and more unique setups entering the zone for 5G, it is opportune to consider a more powerful and adaptable security design for it. Security for synchronous RAN signaling [8] could have a higher level of freedom than asynchronous security aspects. New security outlines with higher adaptability can better address clashes between convenience and security.

7.5 Network Slicing Security

Network slicing not only requires the necessary security from UE accessing the slice but also poses new security challenges. Isolation should be assured for network slices, without which attackers who have access to one slice may launch an attack to other slices. Proper isolation will enable integrity and confidentiality protection. Additionally, it should be ensured that resources of the network infrastructure or a network slice instance are not impacted by another slice instance, to minimize attacks and provide availability [23]. A 5G UE can simultaneously access different network slices for multiple services. Such access can be via various types of radio access networks including both 3GPP and non-3GPP. When the network slice access information is tampered, unauthorized UEs may use such information to establish a connection with the network slice and consume resources.

On the other hand, the advantage of network slicing is that operators can provide tailored security for each slice. Different access authentication and authorization can be provided for tenants of different network slices.

7.6 Vitality Effective Security

While security services such as encryption come with a cost, the cost is no longer an issue for cell phones and comparable gadgets. The vitality cost of encoding one bit is similar to transmitting one bit. In any case, for battery-operated gadgets with a long target lifetime, there might be a need to consider significantly more lightweight arrangements, as each joule devoured could be of significance.

7.7 Cloud Security

Cloud security is a critical concept and will be added to the list of 5G security concerns.

- *Integrity Management*

  Each service provider will have its own integrity management system to control access to data and other assets. Cloud suppliers either incorporate the user's integrity management framework into their own foundation using federation or SSO technology or a biometric-based framework or develop an integrity management arrangement of their own [24]. Cloud ID, for example, gives protection, safeguarding cloud-based biometric distinguishing proof. It connects the private data of users to their biometrics and stores them in a scrambled manner. Making utilization of an accessible encryption method, the biometric distinguishing proof is performed in the scrambled area to ensure that the cloud suppliers or potential aggressors don't access any sensitive information or the contents of individual inquiries.
- *Physical Security*

  Cloud specialist advocate the idea of physically securing the IT hardware (servers, switches, links, and so on.) against unapproved access, obstruction, burglary, fires, surges, and so forth and guaranteeing that fundamental resources such as power are adequately protected to diminish the likelihood of interruption. This is regularly accomplished by serving cloud applications from "world-class" [24] (i.e., professionally conceived, designed, built, overseen, and maintained) server centers.
- *Faculty Security*

  Various information security concerns relating to the IT and other professionals associated with cloud services are typically handled through pre-, para- and post-employment activities such as security screening potential recruits, security awareness and training programs, proactive.

- *Privacy*

  Service providers should guarantee that every single basic datum (credit card numbers, for instance) is properly masked or encoded and that exclusive access are granted to approved users only [24]. Any information that the supplier gathers or delivers about user action in the cloud should also be carefully protected.

8 Security Protocols

Many security protocols have been introduced. We focus mainly on authentication and key exchange protocols for 3GPP network.

8.1 Informal Security Protocols

It is often useful to discuss security protocols informally before proceeding with formal analysis. Therefore, we establish an informal understanding of threat model, security properties, channels, and protocols.

Before going to study about open security protocols, we need to know about Dolev-Yao adversary.

Dolev-Yao Adversary

Large messages are thought to be components of some theoretical variable-based math, and cryptography is a unique activity on that polynomial math [25]. The adversary is thought to be a particular (yet non-deterministic) state machine, and the main route for the enemy to deliver new messages is to play out specific activities on messages it definitely "knows." This model has a to a great degree pleasant component called as straightforwardness. Since all members (genuine and noxious) can be spoken to as state machines, they can be created together to deliver a solitary expansive "framework" machine. Security properties can be communicated as well-being properties about this machine, and such properties can be confirmed consequently. This model likewise has a downside: the Dolev-Yao adversary is entirely frail.

Threat Model

In particular, the Dolev-Yao adversary controls the network, i.e., it can read, intercept, and send messages. Moreover, the adversary can compromise clients, i.e., it can reveal their secrets. Furthermore, the adversary is allowed to apply public functions such as hashing, encryption, or signing on values that she knows. Also, the threat model allows unbounded message lengths, an unlimited number of fresh nonces, and an unlimited number of protocol sessions.

Security Properties

An informal definition of some fundamental security properties is given first and formalized.

Definition (Authenticity) Information is authentic if the original message sender is whom he or she claims to be and the message is unchanged [26].

Definition (Confidentiality, Secrecy) Confidentiality (also called secrecy) is the property of information being protected from disclosure to unauthorized parties.

Definition (Integrity) Information has integrity if it is not modified in any way by unauthorized parties.

Authentication Properties for Protocols

Definition (Aliveness, Informal) A protocol assures to an agent "x" in role X the aliveness of another agent "y" if whenever "x" accomplishes a run of the protocol, apparently with "y" in role Y, then "y" has previously been running the protocol [25].

Definition (Weak Agreement, Informal) A protocol assures to an agent "x" in role X weak agreement with another agent "y" if whenever agent "a" accomplishes "x" run of the protocol, apparently with "y" in role Y, then "y" has previously been running the protocol, apparently with "x" [25].

Definition (Non-injective Agreement, Informal) A protocol assures to an agent "x" in role X non-injective agreement with an agent "y" in role Y on a message M if whenever "x" accomplishes a run of the protocol, apparently with "y" in role Y, then "y" has previously been running the protocol, apparently with "x," and "y" was acting in role Y in his run, and the two agents agree on the message M.

Definition (Injective Agreement, Informal) The injective agreement is defined to be non-injective agreement with the additional property that each run of agent "x" in role X corresponds to a unique run of agent "y" in role Y. The intuitive understanding of injective agreement is that it prevents replay attacks.

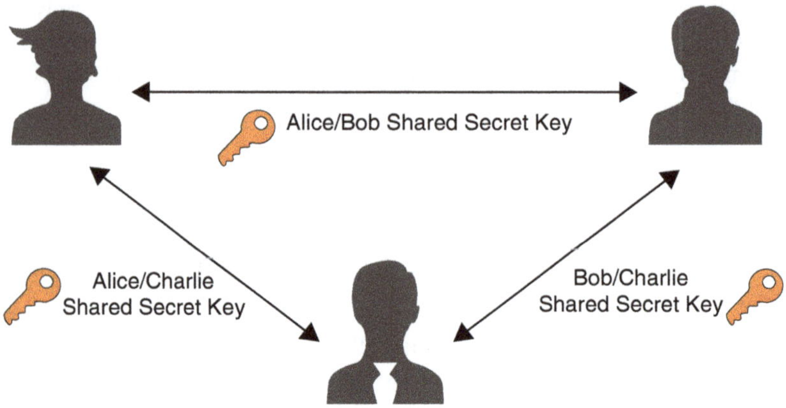

Fig. 4 Alice and Bob message transmission

Channels

For two parties to exchange messages, it is crucial that they be connected in some way.

Definition (Channel) A channel is a logical connection between two parties that can be used to transmit messages.

Recall that the threat model assumes that the adversary controls the network, i.e., it can read and send arbitrary messages over a regular channel. Further, this motivates the following definition of an essential type of channel, making use of the defined notions of informal security properties.

Definition (Secure Channel) A secure channel is a channel that provides confidentiality and authenticity. However, it does not protect messages from being replayed or reordered by the adversary.

Alice and Bob Notation
Protocols will be first specified in an extended form so-called Alice and Bob notation before specifying them formally. Protocols are specified as a set of roles, where every part comprises of a number of steps. Each step sends or receives messages. Moreover, the protocol participants are called agents. Each agent has a name and can execute a protocol in different roles with other agents.

Figure 4 demonstrates the process followed by the Alice and Bob concept. The message which is to be transmitted to the other end is encrypted so that only the person at the other end of will know how to decrypt it regardless of whether it is an insecure communication medium.

In a nutshell, Alice and Bob notation is a compact and succinct description of the messages that the protocol agents exchange in the presence of an attacker. The semantics of the Alice and Bob notation will not be defined rigorously. Instead, the principal conventions are explained, and simple example is given to furnish the understanding.

The following conventions are used:

- If an agent receives a message containing a term *x* and *x* is known to the agent (e.g., because it is the peer's agent name or the agent has sent or received x in an earlier message of the same protocol run), then it must verify that the values of both x's match. This check is implicit in the notation [27].
- If an agent receives a message, it will verify its structure up to the level required to recover all sub-terms it needs, for example, to match the values it knows already or to compute subsequent messages. Also, this is crucial, since the recipient's view may differ from the sender's view.

  For example, if an Alice and Bob protocol specification describes a message (x, hash(y)) for some hash function hash, then an agent who knows only x (and not y) will accept any pair t with x as first element, e.g., t = (x, 0) or t = (x, f ((h, 1))) for some function f.
- Messages marked with an asterisk * are optional, i.e., they can be skipped by both the sending and the receiving agents.
- {|x|}*k* denotes symmetric encryption of the message x with key *k*. The message x can be recovered from {|x|}*k* if and only if an agent knows *k*.
- {m}*sk* denotes the message m signed with key *sk*. A signature in Alice and Bob notation is always hiding, i.e., the message itself cannot be recovered from the signature.
- [[m]]*sk* is an abbreviation for (m,{m}*sk*). It can be understood as a form of non-hiding signature.
- → denotes a secure channel.
- Messages of a protocol are numbered consecutively, starting from 1.

Protocol Example 1

1. A → B: (A, {|n|}*k*)
2. B → A: {B, n}*k*
3. B → C: ({|n|}*k*, *k*)

How the above protocol example runs is informally described below. Assume agent "a" is executing the protocol in role A with an agent "b" in role B. Moreover, "b" is executing the protocol with role A and with an agent "c" in role C.

1. a starts by sending the message (a,{|n|}*k*) over an attacker-controlled channel to b. Here, n is to be understood as a fresh nonce and *k* is a long-term key shared between a and b. When b receives the message, it verifies the value of a in the message and tries to decrypt the value n.
2. If b has accepted the first message from a, it replies with the message {b, n}*k* over an attacker-controlled channel. When a receives the message, it verifies the value of band n as well as the signature.
3. After b has sent the second message, it can optionally send the message ({n}*k*, *k*) to cover a secure channel. c will accept the message in any case, even if it has

the wrong structure (e.g., if it is a constant string). Further, this is because c does not know the key *k* or the term {n}*k* and it does not need to extract one of those terms to compute a subsequent message.

Attack Scenarios

The term attack scenario is used to outline protocol attacks. An attack scenario uses the same notation as Alice and Bob protocols. Additionally, we use the following conventions.

- We write Adv(R) to denote the adversary masquerading as agent R.
- Parallel protocol instances are indented in case an attack requires multiple instances.
- Messages that are not relevant to the attack are omitted.
- We distinguish different runs of the same agent A (if there is more than a single run) with indices A [0], A [1]...

Assume that in the protocol example, B would claim injective agreement with A on n after it received the first message. It is easy to see that the property is violated because the adversary can masquerade as A and resend the first message to B.

Attack Scenario 1 outlines the attack.

Attack Scenario 1 (Example)

1. A → B [0]: (A, {|n|}*k*)
2. Adv (A) → B [1]: (A, {|n|}*k*)

The outlined attack violates injective agreement of B with A on the value n. It is a trivial replay attack that makes B accept the same value n twice.

9 Channel Security

Physical layer security achieves information confidentiality based on data theoretic methodologies and has got significant progress. The key idea behind physical layer security is to utilize the normal haphazardness of the transmission channel to guarantee security in the physical layer. The move toward 5G mobile communication poses new challenges for physical layer security to look into.

9.1 Introduction

These days, mobile communication systems have been broadly utilized in regular citizen and military applications and have become a vital piece of our day-to-day life. Individuals depend intensely on the systems for transmission of vital/private data, for example, credit card data, e-well-being information, and control messages.

As a result, security is a key issue for future 5G mobile networks. Normally, existing security measures depend on bit-level cryptographic techniques and associated protocols at different levels of the data processing stack. These solutions have several disadvantages. First, standardized protections within public wireless networks are not safe enough, and many of their weaknesses are well known. Second, even if enhanced ciphering and authentication protocols exist, they incur heavy constraints and high additional costs for the users of public networks. Therefore, new security approaches are rooted in information theory fundamentals and focus on the secrecy capacity of the propagation channel [28]. This is referred to as physical layer security.

The advantages of employing physical layer security techniques for 5G networks compared to that of cryptography techniques are twofolds.

- Initially, physical layer security systems don't depend on computational intricacy [29].
- Second, the structures of 5G networks are usually decentralized, which implies devices may randomly connect to or exit the network at any time instant.

As a result, physical layer security techniques can be used to either perform secure data transmission directly or generate the distribution of cryptography keys in the 5G networks. Specifically, we center on the following innovations.

9.2 *Physical Layer Security Coding*

Although the first physical layer security code appeared around 1970s, the design of specific security codes which can be used in practical communication systems is still challenging. We survey the best in class of three crucial physical layer security codes, including low-density parity check (LDPC) codes, polar codes, and lattice codes [30].

LDPC Codes

The LDPC codes are the first secrecy-capacity-achieving codes regarding weak secrecy. LDPC codes have been intended for the Gaussian wiretap channel. The physical layer security communication is fulfilled by punctured LDPC codes under the metric of bit error rate (BER), where the secrecy data bits are covered up in the punctured bits. In this way, these data bits are not transmitted through the channel but rather can be decoded at the recipient side in light of the non-punctured piece of the codeword.

This coding scheme can yield a BER close to 0.5 at the eavesdropper's (Eve Channel) side while significantly decreasing the security gap compared to the non-punctured LDPC codes. However, the punctured LDPC codes result in higher transmit power compared to the non-punctured LDPC codes. Further, to solve this problem, a nonsystematic coded transmission design by scrambling the information bits has been proposed. This scrambling technique achieves a security level comparable to the design based on puncturing without expanding the transmit control. This scrambling configuration has been applied to parallel Rayleigh dispersed channels by exploiting the equivocation rate of eavesdropper's channel as an optimization criterion. A summary of the attributes of LDPC codes is given in Table 1. 5G enhanced mobile broadband (eMBB) use case will adopt LDPC codes for channel coding on the data channel.

Polar Codes

For the weak secrecy criterion, a polar coding scheme is constructed to attain the secrecy capacity for the symmetric binary-input memoryless wiretap channel under the condition that the channel of the eavesdropper is degraded relative to the main channel of the desired user [31]. The main idea is to select only those bit channels which are suitable for both the desired user and the eavesdropper to transmit random bits. Moreover, [31] those bit channels which are suitable for the desired user but bad for the eavesdropper are used to transmit information bits.

This coding scheme is applied to a critical agreement problem over the block fading wiretap channel. The safe polar code is utilized for each fading block, from which the secrecy keys are produced in light of standard privacy amplification techniques [32].

Also, this coding scheme is extended to the multiple access wiretap channel (MA-WC), the broadcast channel with confidential message (BC-CM), and the interference channel with confidential message (IC-CM).

This coding scheme has been applied to accomplish the Shannon capacity of discrete memoryless BC-CM [32]. A summary of the attributes of polar codes is given in Table 2. 5G eMBB use case will adopt polar codes for channel coding on the control channel.

Table 1 LDPC codes for physical layer security

Main channel	Eve channel	Criterion	Constituent codes
Noiseless	BEC	Weak secrecy	Duals of LDPC
BEC	BEC	Weak secrecy	Two-edge LDPC
Noiseless	BEC	Strong secrecy	Duals of LDPC
Gaussian	Gaussian	BER	Punctured LDPC
Gaussian	Gaussian	BER	Non-punctured LDPC
Parallel Rayleigh	Parallel Rayleigh	BER	Non-punctured LDPC
Gaussian	Gaussian	Equivocation rate of Eve	Irregular LDPC

Table 2 Polar codes for physical layer security

Channel	Criterion	Main contribution
Symmetric binary-input memory less degraded wiretap channel	Weak secrecy	Achieve secrecy capacity
Symmetric binary-input memory less degraded wiretap channel	Weak secrecy	Achieve rate-equivocation region
Symmetric binary-input memory less degraded wiretap channel	Weak secrecy	Generate a key agreement
General wiretap channel MA-WC, BC-CM, IC-CM	Weak secrecy	Achieve secrecy capacity Achieve secrecy rate regions
Deterministic wiretap channel	Weak secrecy	Achieve secrecy capacity
Bidirectional relay networks with confidential messages	Weak secrecy	Achieve capacity-equivocation region
Symmetric binary-input memory less degraded wiretap channel	Strong secrecy	Achieves both security and reliability
General wiretap channel	Strong secrecy	Achieve secrecy capacity
Discrete memory less BC-CM	Strong secrecy	Achieve secrecy capacity

Table 3 Lattice codes for physical layer security

Channel	Criterion	Main contribution
Gaussian wiretap channel	Secrecy gain	Define secrecy gain
Gaussian wiretap channel	Secrecy gain	Propose a method to examine the secrecy gain
Gaussian wiretap channel	Secrecy gain	Construct best lattice codes for dimensions $8 < n \leq 23$
Rayleigh wiretap channel	Secrecy gain	Construct a wiretap lattice code
Gaussian wiretap channel	Weak secrecy	Construct a wiretap lattice code
Gaussian wiretap channel	Strong secrecy	Design wiretap lattice codes
Gaussian BC with a confidential message	Strong secrecy	Propose a superposition lattice code

Lattice Codes

For wiretap lattice codes, a notation of secrecy gain is defined, which reflects the eavesdropper's correct decoding probability [32]. Asymptotic analysis of the secrecy gain shows that it scales with the dimension of the lattice. A summary of the attributes of lattice codes is given in Table 3.

Other lattice code designs for the wiretap channel include nested lattice code for cooperative jamming, interference channels, and Gaussian relay networks.

Table 4 Secure massive MIMO with passive eavesdropper

System model	Main contribution
Multi-cell multiuser, one desired user, one eavesdropper	Matched filtering precoding and AN generation designs
Multi-cell multiuser, one desired user, one eavesdropper	Regularized channel inversion and AN generation designs
One desired user, multiple eavesdroppers	AN-aided secure transmission designs
Single-cell, multiple desired users, one eavesdropper	Distributed power allocation under security constraints
One desired user, one eavesdropper	Secure transmission with finite alphabet inputs
Relay-aided, one desired user, one eavesdropper	Secrecy performance analysis and power allocation designs

9.3 Massive MIMO

Deploying large antenna arrays significantly increases the capacity of channels. Massive MIMO is a promising technology for effective transmission of massive data and is viewed as one of the "big three" 5G technologies.

Passive Eavesdropper Scenarios

Physical layer security for massive MIMO systems with passive eavesdroppers has been newly studied [33]. The impact of multi-cell interference and pilot contamination on the achievable ergodic secrecy rate are analyzed, and several matched filtering, precoding and artificial noise (AN) generation designs are proposed to mitigate the eavesdropper's channel [33] and protect the desired user's channel.

For single-cell multiuser massive MIMO systems with distributed antennas, three security-constrained power allocation schemes are designed by maximizing the minimum user's signal-to-interference-noise ratio (SINR) subject to the eavesdropper's SINR and the sum power constraint and reducing the sum transmit power subject to SINR [34] constraints of the users and the eavesdropper, respectively.

A summary of secure massive MIMO systems with passive eavesdropper(s) is given in Table 4.

Other secure massive MIMO works with passive eavesdroppers include a secure transmission for massive MIMO systems with limited radio-frequency and hardware impairments, secure strategies in the existence of a massive MIMO eavesdropper, secrecy outage probability analysis for a massive MIMO system, and so on [35].

Active Eavesdropper Scenarios

Most physical layer security research work assumes that the perfect channel knowledge of the appropriate user is available at the transmitter and won't delve into details of the procedure required to obtain this channel knowledge [31]. In time duplex division (TDD) communication frameworks, the users in an uplink

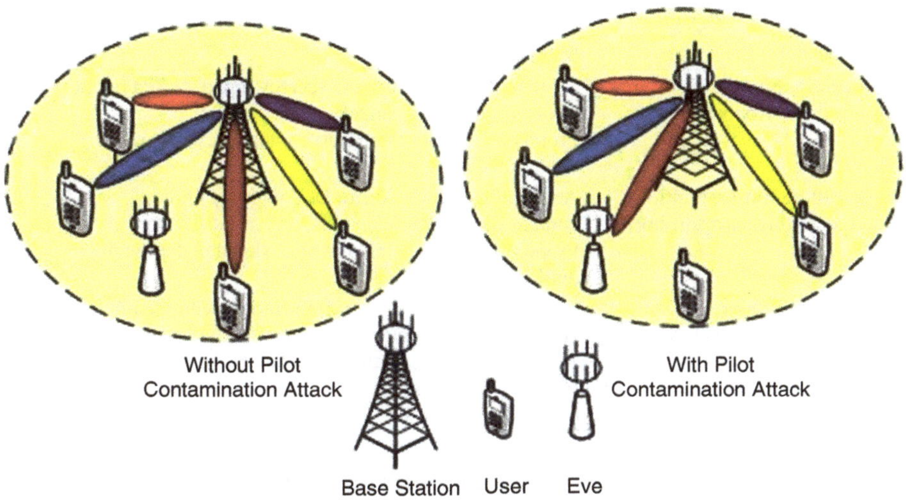

Fig. 5 Secure massive MIMO with active eavesdropper

preparing stage will send pilot signals to the base station (BS) to assess the channel for the ensuing downlink transmission. From the eavesdropper's perspective, it can effectively send a similar pilot signal as the clients to attack this uplink channel training stage and thus significantly increase its eavesdropping capability.

This pilot contamination attack causes a severe secrecy threat to TDD-based massive MIMO systems [32]. On one hand, large antenna arrays beam-forming leads to the hardening of the channel, which prevents the exploitation of channel fluctuations caused by fading to sharpen the secrecy performance. On the other hand, as illustrated in Fig. 5, the pilot contamination [36] attack causes the transmitter to beam form toward the eavesdropper instead of the desired user. If the eavesdropper's pilot power is sufficiently large, a desired secrecy rate may not be achievable. Systematical analysis of the secrecy threat caused by the pilot contamination attack for multi-cell multiuser massive MIMO systems over correlated fading channels has been performed [32]. Then, a matched filter precoding and AN generation design and a null space design are provided to nullify the pilot contamination attack for weakly correlated channels and highly correlated channels, respectively. A unified design which combines the matched filter precoding and AN generation design and null space design is also proposed. MIMO systems are the same as the maximum DoF (degrees of freedom) of massive MIMO systems when the eavesdropper does not exist. However, if the pilot contamination attack exists, the maximum secure DoF of massive MIMO systems could be zero. An estimator is designed at the BS side to evaluate the leakage [37].

Then, the BS and the desired user perform the reliable, secure communication by adjusting the lengths of the secret key based on the estimated information leakage. A summary of secure massive MIMO systems with active eavesdropper(s) is given in Table 5.

Table 5 Secure massive MIMO with active eavesdropper

System model	Main contribution
Multi-cell multiuser, one desired user one eavesdropper	Systematically analyze the secrecy threat caused by the pilot contamination attack Propose efficient schemes to combat the pilot contamination attack
Single-cell multiuser, multiple desired users one eavesdropper	Analyze maximum secure DoF with the pilot contamination attack Propose a plan to shield the pilot signal under the pilot defilement attack
Single-cell multiuser, one desired user one eavesdropper	Employ secret key agreement protocol with the pilot contamination attack Adjust the lengths of the secret key based on the estimated information leakage

9.4 Millimeter Wave (mmWave) Communications

Abundant spectra within the high-frequency band may result in significantly different propagation environments for physical layer secure communication. To understand mmWave secure transmission more clearly, research works for both point-to-point and network mmWave communication systems are introduced.

One of the most promising potential 5G technologies under consideration is the use of high-frequency signals in the millimeter-wave frequency band that could allocate greater bandwidth to deliver faster, higher-quality video and multimedia contents [38]. Compared to microwave networks, the mmWave networks have various new characteristics such as a large number of antennas, short range and highly directional transmissions, different propagation laws, sensitive to blockage effects, and so on. Therefore, secure mmWave communications will be different from conventional secure microwave communications.

The mmWave communication system is usually equipped with a larger number of antennas at the transmitter with a limited number of radio-frequency (RF) chains [39]. To take advantage of this point, let us consider another approach by using an antenna subset modulation (ASM) technique to reach secure mmWave communication at the physical layer. The proposed approach utilizes a subset of the antenna array to formulate a directional modulation signal intended for the desired user. By randomly choosing the antenna subset for each symbol, the received signal for the undesired user becomes a randomized noise. Therefore, secure transmission is achieved. This ASM technique can be further extended to mmWave vehicular communication systems.

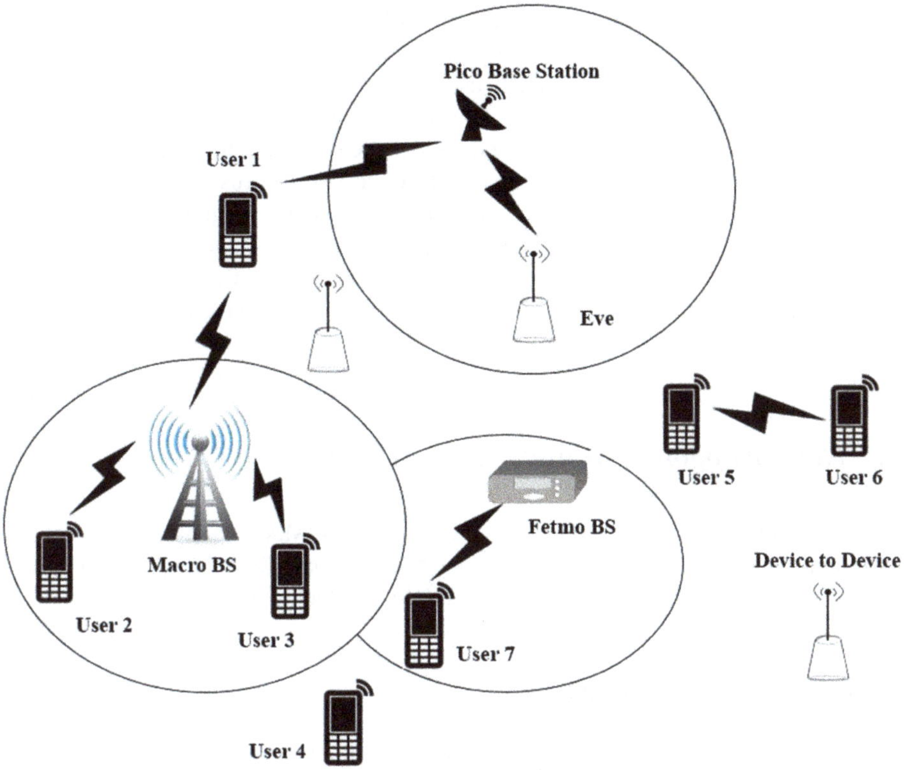

Fig. 6 A four-tier macro-/pico-/femto-/D2D heterogeneous network with users and eavesdroppers

9.5 Heterogeneous Networks

In general, a heterogeneous network is composed of various tiers of networks which operate in the same system bandwidth [37]. We describe in detail on how to design transmission schemes to secure multitier communications.

Physical Layer Security in Heterogeneous Networks

The 5G heterogeneous systems ought to insightfully and consistently incorporate different nodes to frame a multitier hierarchical architecture [40], including the macro-cell tiers with high-power nodes for extensive radio coverage regions, the small cell tiers with for small radio areas, and the gadget levels which bolster gadget-to-gadget interchanges. Figure 6 shows a typical four-tier macro-/pico-/femto-/D2D heterogeneous network with users and eavesdroppers.

This multitier architecture brings new challenges to the investigation of physical layer security compared to the conventional single-tier topology [41]. For example, the locations of the high-/low-power nodes will have a significant impact on the physical layer security design, which needs to be modeled and analyzed correctly. The optimal selection policy for each user among high-/low-power nodes under security constraints becomes difficult. The protection of confidential and privacy data between connected devices against data leakage requires sophisticated designs. Moreover, heterogeneous networks may introduce severe cross-tier interference. These aspects should be taken into consideration when designing reliable and secure data transmission schemes. Also, [40] users are accessible to [alternatively, should have access to] an arbitrary tier, e.g., open access. Therefore, particular user association policies that coordinate both quality of service and secrecy are necessary.

9.6 Non-orthogonal Multiple Access (NOMA)

As a multiple access technology, the security of NOMA communications is a paramount concern to which more attention should be paid [39]. The physical layer security technology can be combined with NOMA to tackle this issue.

Physical Layer Security of NOMA

NOMA plays a crucial role in providing substantial system throughput, high reliability, improved coverage, low latency, and massive connectivity in 5G wireless networks. As a result, NOMA has been recognized as an essential enabling technology in 5G wireless communication systems [42]. Because of the spectral efficiency benefit, NOMA has been newly incorporated into 3GPP Long-Term Evolution Advanced (LTE-A), which additionally proves the significance of NOMA in future communication systems. Consequently, giving an unrivalled level of security for NOMA innovation is one of the most urgent needs in the outline and execution of the 5G communication systems. Significant research is needed to efficiently consolidate physical layer security with NOMA [39]. However, a few difficulties should be addressed in the design process, for example, the different transmit powers and heterogeneous security requirements of users. Additionally, participation of clients offers an intriguing alternative to improve the secrecy execution.

In Fig. 7, an eavesdropper-exclusion zone is established. To reduce the SIC (successive interference cancellation) complexity at the receiver, a user paring scheme is employed, where one user in the internal zone and one user in the external zone are allocated in the same resource slot. When the base station only has a single antenna, the secrecy outage probability is analyzed.

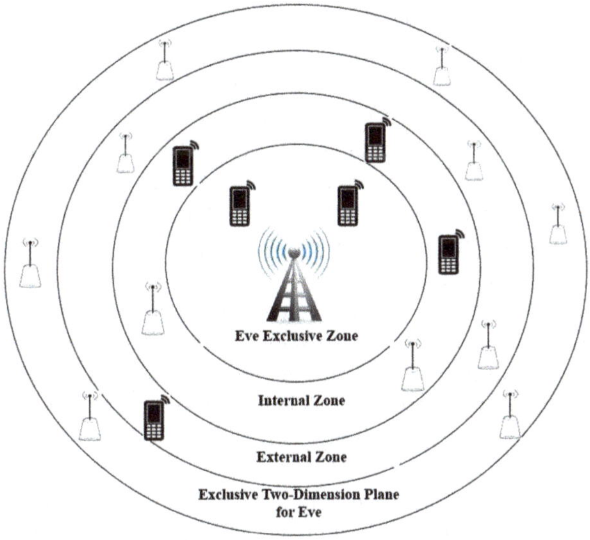

Fig. 7 Network model for secure NOMA transmission

9.7 *Full Duplex Technology*

Full duplex technology brings both opportunity and challenge for the physical layer security communication. On one hand, the full duplex innovation empowers the recipient to produce extra AN to interfere the eavesdropper. Then again, the eavesdropper with full duplex innovation can effectively attack the communication procedure while eavesdropping. In general, we talk about four classes of full duplex physical layer security interchanges, including the full duplex receiver, the full duplex transmitter and receiver, the full duplex base station, and the full duplex eavesdropper.

Full Duplex Receiver

Firstly, a single-antenna transmitter, a two-antenna full duplex receiver, and a single-antenna eavesdropper wiretap channel were studied [43], where the full duplex receiver uses one antenna to receive the signal and another antenna to send AN to the eavesdropper. Perfect self-interference cancellation (SIC) is assumed at the receiver.

The closed-form expression of the secrecy outage probability for the transmission scheme proposed in [44] is derived to investigate the joint transmit and receive beam-forming design for a single-antenna input, multiple-antenna output, and multiple-antenna eavesdropper (SIMOME) wiretap channel with imperfect SIC [45]. The full duplex receiver transmits AN to the eavesdropper while accepting data from the transmitter. For the global perfect CSI assumption, the linear receiver matrix and the AN generation matrix which boost the achievable secrecy rate are mutually designed. It is demonstrated that unlike the half duplex case, the secrecy rate no longer saturates at high SNR for the full duplex case.

Further, a closed-form expression is derived for the maximal achievable secure degrees of freedom of the MIMO ME (multiple-antenna eavesdropper) wiretap channel with a full duplex receiver under the global perfect CSI and perfect SIC assumptions [43]. Both the transmitter and the full duplex receiver will send AN to debase the channels of the eavesdroppers. For this situation, the precoding matrix and the AN generation matrix are streamlined mutually to boost the achievable secrecy rate. The secure communication in a single-input-single-output multiple-antenna eavesdropper (SISOME) wireless ad hoc network is analyzed where a hybrid full/half duplex receiver deployment strategy is employed [46]. The fractions of full duplex receivers which optimize the secure link number, the network-wide secrecy throughout, and the network-wide secrecy energy efficiency are derived.

Secure bidirectional communication is investigated where two multiple-antenna full duplex nodes communicate with each other in the presence of multiple-antenna eavesdroppers. Global perfect CSI and imperfect SIC assumptions are adopted. The beam-forming vectors are designed to reduce the total transmit power subject to the constraints of secrecy and QoS parameters. Assuming global perfect CSI and perfect SIC, the secrecy sum rate of bidirectional full duplex communication systems is maximized within the sight of a single-antenna eavesdropper under the sum transmit power constraint [46]. A null space-based suboptimal design is also proposed to reduce the computational complexity.

Full Duplex Base Station

Consider a multiple-antenna full duplex base station which communicates with a single-antenna transmitter and a single-antenna receiver simultaneously with single-antenna eavesdropper [47] (Fig. 8). The joint precoding and AN generation design at the base station with global perfect CSI and perfect SIC is investigated to guarantee both the uplink and downlink transmission security. It is expected that exclusive defective [48] CSIs of eavesdroppers are accessible to the base station. A vigorous asset designation is intended to limit the total of uplink and downlink transmit power subject to the uplink and downlink data rate and security rate requirements. As illustrated in Fig. 9, the proposed configuration accomplishes significantly higher-power efficiency compared to the baseline ZFBF (zero-forcing beam-forming) scheme.

Full Duplex Eavesdropper

Consider a multiple-antenna full duplex active eavesdropper which simultaneously eavesdrops and attacks the legitimate MIMO communication link. It is expected that

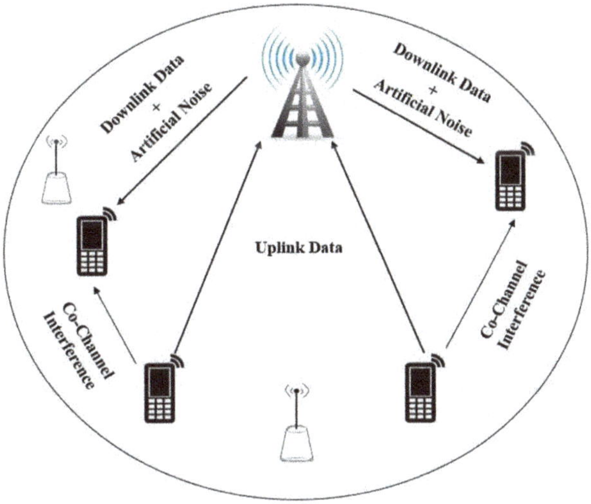

Fig. 8 Secure communication for the full duplex base station network

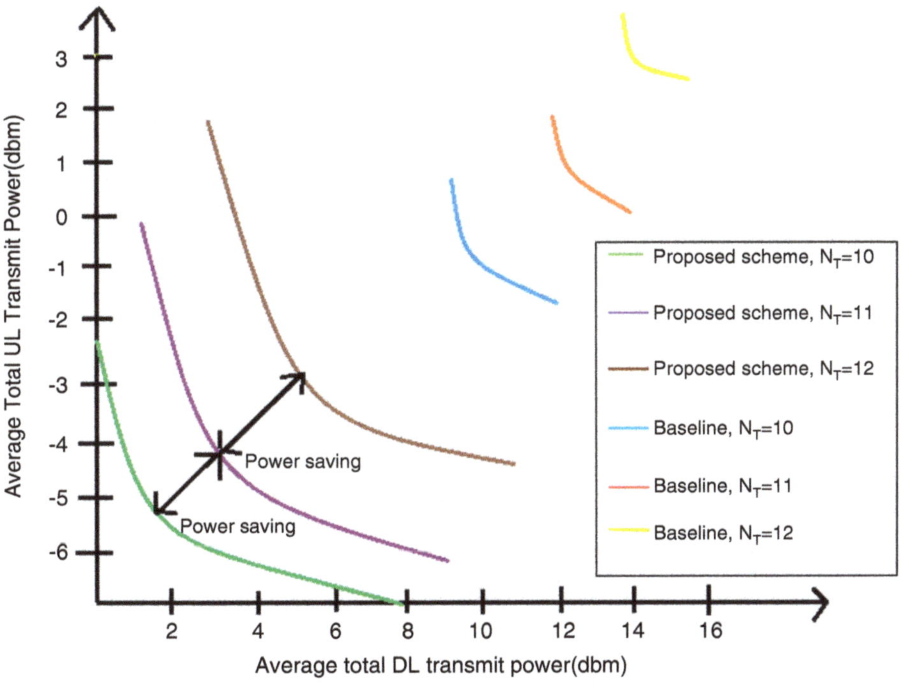

Fig. 9 Tradeoff between the downlink and uplink total transmit powers

the eavesdropper has information of the channels among all nodes and imperfect estimation of the self-interference channel. The jamming signals which minimize the secrecy rate are designed based on the Karush-Kuhn-Tucker (KKT) analysis.

Then, the optimal transmission strategies at both the eavesdropper and the legitimate user's sides are designed.

References

1. G. Asvin, H. Modi, S.K. Patel, 5G technology of mobile communication: A survey, in *IEEE International Conference on Intelligent Systems and Signal Processing (ISSP)*, Gujarat, 2013, pp. 288–292
2. Internet source – isindexing.com
3. M.K. Arjmandi, 5G overview: Key technologies, in *Opportunities in 5G Networks: A Research and Development Perspective*, ed. by F. Hu (CRC Press, Boca Raton, 2016)
4. Internet source – www.nou.edu.ng
5. P. Sharma, Evolution of mobile wireless communication networks-1G to 5G as well as future prospective of next generation. Int. J. Comput. Sci. Mob. Comput. **2**(8), 47–53 (2013)
6. Internet source – scm-fallersleben.ciando.com
7. M. Chen, Y. Qian, S. Mao, W. Tang, X. Yang, Software defined mobile network security. Mob Netw Appl **21**(5), 729–743 (2016)
8. Internet source – www.ericsson.com
9. M. Liyanage, I. Ahmad, A. B. Abro, A. Gurtov, M. Ylianttila (eds.), *A Comprehensive Guide to 5G Security* (Wiley, Hoboken, 2018)
10. V.-G. Nguyen, A. Brunstrom, K.-J. Grinnemo, J. Taheri, 5G mobile networks -requirements, enabling technologies, and research activities, in *A Comprehensive Guide to 5G Security*, (Wiley). https://www.etsi.org/etsi-security-week-2017/5g-security
11. W. El-Beaino, A.M. El-Hajj, Z. Dawy, On radio network planning for next generation 5G networks: A case study, in *International Conference on Communications, Signal Processing, and Their Applications (ICCSPA)*, Sharjah, 2015, pp. 1–6. https://doi.org/10.1109/ICCSPA.2015.7081315
12. Y. Elias, Z. Dawy, LTE radio network planning with Het-Nets: BS placement optimization using simulated annealing, in *MELECON 2014–2014 17th IEEE Mediterranean Electro-Technical Conference*, 2014
13. C.Y. Lee, H.G. Kang, Cell planning with capacity expansion in mobile communications: A Tabu search approach. IEEE Trans. Veh. Technol. **49**(5), 1678–1691 (2000)
14. Internet Source – hal.inria.fr
15. Z. Ribeiro, L.A. DaSilva, A framework for the dimensioning of broadband mobile networks supporting wireless internet services. IEEE Wirel. Commun. **9**(3), 6–13 (2002)
16. K. Tutschku, P. Tran-Gia, Spatial traffic estimation and characterization for mobile communication network design. IEEE J. Sel. Areas Commun. **16**(5), 804–811 (1998)
17. R. Pattuelli, V. Zingarelli, Precision of the estimation of area coverage by planning tools in cellular systems. IEEE Pers. Commun. **7**(3), 50–53 (2000)
18. A. Taufique et al., Planning wireless cellular networks of future: Outlook, challenges and opportunities. IEEE Access **5**, 4821–4845 (2017)
19. Internet source – 5g.ieee.org
20. Internet source – pubs.sciepub.com
21. L. Hermann, MIMO OFDM space time coding – spatial multiplexing, increasing performance and spectral efficiency in wireless systems, Part I technical basis (Technical report), 2007
22. A. Agarwal, G. Misra, K. Agarwal, The 5th generation mobile wireless networks – key concepts, network architecture and challenges. Am. J Electr Electron Eng **3**(2), 22–28 (2015)

23. Xiaowei Zhang Detecon International GmbH, Cologne, Germany, Andreas Kunz Lenovo, Oberursel, Germany, Stefan Schröder T-Systems International GmbH, Bonn, Germany. 2017 IEEE Conference on Standards for Communications and Networking (CSCN)-Overview of 5G security in 3GPP (2017).
24. Internet Source – www.nosmut.com
25. Internet source – tamarin-prover.github.io
26. B. Schmidt, Formal Analysis of Key Exchange Protocols and Physical Protocols, PhD Thesis, ETH Zurich, 2012
27. S. Meier, Advancing Automated Security Protocol Verification, PhD Thesis, ETH Zurich, 2013
28. Internet source – www.phylaws-ict.org
29. N. Yang, L. Wang, G. Geraci, M. Elkashlan, J. Yuan, M. Di Renzo, Safeguarding 5G wireless communication networks using physical layer security. IEEE Commun. Mag. **53**, 20 (2015)
30. Internet source – www.faqs.org
31. M. Baldi, G. Ricciutelli, N. Maturo, F. Chiaraluce, Performance assessment and design of finite length LDPC codes for the Gaussian wiretap channel, in *Proc. Int. Conf. Commun. (ICC'2015)*, London, June 2015, pp. 435–440
32. H. Mahdavifar, A. Vardy, Achieving the secrecy capacity of wiretap channels using polar codes. IEEE Trans. Inf. Theory **57**, 6428–6443 (2011)
33. Y. Wu, R. Schober, D.W.K. Ng, C. Xiao, G. Caire, Secure massive MIMO transmission with an active eavesdropper. IEEE Trans. Inf. Theory, 1–22 (2016)
34. K. Guo, Y. Guo, G. Ascheid, Security constrained power allocation in MU-massive MIMO with distributed antennas. IEEE Trans. Wireless Commun. **15**(12), 8139–8153 (2016)
35. Y. Zhang, A. Liu, C. Gong, G. Yang, S. Yang, Polar-LDPC concatenated coding for the AWGN wiretap channel. IEEE Commun. Lett. **18**, 1683–1686 (2014)
36. Y. Wu, R. Schober, D.W.K. Ng, C. Xiao, G. Caire, Secure massive MIMO transmission in the presence of an active eavesdropper, in *2015 IEEE International Conference on Communications (ICC)*, 2015
37. T.T. Do, H.Q. Ngo, T.Q. Duong, T.J. Oechtering, M. Skoglund, Massive MIMO pilot retransmission strategies for robustification against jamming. IEEE Wireless Commun. Lett. **6**(1), 58–61 (2017)
38. Internet source – www.scientificamerican.com
39. Y. Wu, A. Khisti, C. Xiao, G. Caire, K.-K. Wong, X. Gao, A survey of physical layer security techniques for 5G wireless networks and challenges. IEEE J Sel. Areas Commun. **36**(4), 679–695 (2018)
40. D.B. Rawat, K. Neupane, M. Song, A novel algorithm for secrecy rate analysis in massive MIMO system with target SINR requirement, in *Proc. INFOCOM'2016*, San Francisco, April 2016, pp. 1–6
41. H.-M. Wang, T.-X. Zheng, *Physical Layer Security in Random Cellular Networks* (Springer, Singapore, 2016)
42. Z. Ding, Z. Zhao, M. Peng, H.V. Poor, On the spectral efficiency and security enhancements of NOMA assisted multicast-unicast streaming. IEEE Trans. Commun. **65**(7), 3151–3163 (2017)
43. G. Zheng, I. Krikidis, J. Li, A.P. Petropulu, B. Ottersten, Improving physical layer secrecy using full-duplex jamming receivers. IEEE Trans. Signal Process. **61**, 4962–4974 (2013)
44. Wireless Networks, Springer, 2016. https://www.cs.cmu.edu/~prs/wirelessS16/
45. G. Chen, Y. Gong, P. Xiao, J.A. Chambers, Physical layer network security in the full- duplex relay system. IEEE Trans. Inf. Forensics Secur. **10**(3), 574–583 (2015)
46. L. Li, Z. Chen, D. Zhang, J. Fang, A full-duplex Bob in the MIMO Gaussian wiretap channel: Scheme and performance. IEEE Signal Process Lett. **23**, 107–111 (2016)
47. Z.H. Awan, A. Zaidi, L. Vandendorpe, Multi access channel with partially cooperating encoders and security constraints. IEEE Trans. Inf. Forensics Secur. **8**, 1243 (2013)
48. F. Zhu, F. Gao, M. Yao, H. Zou, Joint information and jamming-beam forming for physical layer security with full duplex base station. IEEE Trans. Signal Process. **64**(12), 6391–6401 (2014)

Poorna Pravallika Sriram is a B.Tech graduate in the Department of Electronics and Communication Engineering, Vel Tech Rangarajan Dr. Sagunthala R&D Institute of Science and Technology. She has 3 years of work experience on IoT (Internet of Things) as she had a heed interest toward it. She had published papers in national and international conferences on PCB fabrication, image processing, and IoT domains. She had worked as a research analyst at National Ilan University (NIU) on Narrowband Internet of Things (NB-IoT) as a TEEP (Taiwan Education Experience Program) student.

Hwang-Cheng Wang received the M.S. degree in Computer Science and Ph.D. degree in Electrical Engineering, both from the University of Southern California, Los Angeles, California, USA. He is currently a professor in the Department of Electronic Engineering, National Ilan University (NIU), Taiwan. He is in charge of the Embedded Systems and Mobile Computing Laboratory at NIU. Prior to joining NIU, he had served at the Data Communication Institute, Ministry of Transportation and Communications, as a senior technician. He was a visiting scholar at the Next Generation Internet Research Center, Beijing Jiaotong University and Information Science Institute, Academia Sinica. His research interests are in wireless networks, mobile communications, RFID, embedded systems, and innovative combination of information and communication technologies.

Hema Ganesh Jami is from Andhra Pradesh, India. He studied Bachelor of Technology in Electronics and Communication Engineering at Vel Tech University and started master's at National Tsing Hua University, Taiwan, in Communication Engineering. He got selected for research internship at National Ilan University by Taiwan Experience Education Program in January 2018. He has done his internship on Narrowband IoT, with the guidance given by Prof. H.C Wang.

Kathiravan Srinivasan received his B.E., in Electronics and Communication Engineering and M.E., in Communication Systems Engineering from Anna University, Chennai, India. He also received his Ph.D., in Information and Communication Engineering from Anna University Chennai, India. He is presently working as an associate professor in the School of Information Technology and Engineering at Vellore Institute of Technology (VIT), India. He was previously working as a faculty in the Department of Computer Science and Information Engineering and also as the deputy director – Office of International Affairs at National Ilan University, Taiwan. He has won the Best Conference Paper Award at 2018 IEEE International Conference on Applied System Innovation, Chiba, Tokyo, April 13–17, 2018. Further, he has also received the Best Service Award, Department of Computer Science and Information Engineering, National Ilan University, Taiwan. In 2017, he has won Best Paper Award at 2017 IEEE International Conference on Applied System Innovation, Sapporo, Japan, May 13–17, 2017, and Best Paper Award at International Conference on Communication, Management and Information Technology (ICCMIT 2017), Warsaw, Poland. In 2016, he received the Best Service Award as the Deputy Director at Office of International Affairs, National Ilan University. He is presently serving as the associate editor for IEEE Access and Editorial Board Member and reviewer for various SCI, SCIE, and Scopus indexed journals. He has played an active role in organizing several international conferences, seminars, and lectures. He has been a key note speaker in many international conferences and IEEE events.

5G Applications and Architectures

Dinesh Kumar Sah, D. Praveen Kumar, Chaya Shivalingagowda, and P. V. Y. Jayasree

Brief Summary

Considering the present research direction of the Fifth Generation (5G) network one can only draw an observation regarding the requirements for effective protocols, algorithms, and framework for a steamroller transition. The objective of the chapter is to sketch the future adaptation of an application eagerly awaiting a high data-rate to carry out those objectives with the development of protocols or openwork to resolve the use and security threats. In this chapter, the focus is on the application of 5G and on the techniques that are likely to adapt to serve it. It begins with a brief introduction to proper applications and promising methods. We cover a great deal of promising technology, such as non-orthogonal multiple access (NOMA), massive multiple input and multiple output (MIMO), and millimeter wave, with an overview of various architectures, such as cross-layer designs. Moreover, the importance and an overview of software-defined networking/network function visualization (SDN-NFV)-based models are explained, which is helping greatly with the transition. The significance of the work is also enhanced because of the research gap in the current proposal, such as the existing empirical model presented for 5G. Apart from specific issues with the existing proposal, some general parameters (time, space, and hardware resources) and their consequences are also listed. The limitations posed

D. K. Sah (✉) · D. P. Kumar
Indian Institute of Technology (ISM), Dhanbad, Jharkhand, India

C. Shivalingagowda
Kalsekar Engineering College, New Panvel Mumbai and GITAM University, Vizag, Andhra Pradesh, India

P. V. Y. Jayasree
GITAM University, Vizag, Andhra Pradesh, India
e-mail: pvyjayasree@gitam.edu

D. N. K. Jayakody et al. (eds.), *5G Enabled Secure Wireless Networks*,
https://doi.org/10.1007/978-3-030-03508-2_2

by the exciting technology have been given individually in the chapter, where the proposal has been described. This chapter provides insights into the evolution of cellular technology and can be used as a guideline for developing technology toward 5G. Finally, we conclude our chapter with a good insight into industry initiatives and academic collaboration.

1 Brief Introduction to 5G

In the world of wireless communication 5G is revolutionary and procures communications with a high spectral efficiency, a high peak rate, low end-to-end latency, high reliability, significant improvement in QoS, and multiple device connectivity [1–3]. It provides a maximum peak data rate (PDR) of 10 to 20 Gbps per user under certain conditions, which is feasible compared with existing technology. The round trip latency is reduced ten-fold compared with existing 4G technology, i.e., $1\,ms$. 5G technology is energy efficient and portable for connecting with other wireless technologies such as 2G, 3G, 4G, Wi-Fi, WPAN. For efficient communication, 5G uses higher frequencies in the network [4]. Although the previous chapter provided a broad introduction, in this chapter, the perception of the architectural need is established.

Wireless and telecommunication infrastructure has grown exponentially after 'the 2000s', although the load on the network is growing exponentially. The estimated cost of infrastructure is a trillion dollars, which poses a challenge in the adoption of any newly emerging technology. To reduce this significance cost, the adoption of hardware seems infeasible and searches for software-defined infrastructure have become the impetus for now-or-never transformation. Research is underway to make the transformation to 5G and to reduce the cost of adopting the new technology. Therefore, once the 4G is in its deployment, many architectural designs were being ready for upcoming 5G adoption. Various forces drive network architecture transformation [5]:

a. Complex networks incorporating multiple services, standards, and site types
b. Coordination of multi-connectivity technologies
c. On-demand deployment of service anchors
d. Flexible orchestration of network functions
e. Shorter period of service deployment

2 Applications

The number of devices connected through a wireless network has grown exponentially. The advancement of massive MIMO is playing a key role in the development of low-cost sensor devices, opening up the paradigm of newer applications. Although empirical application in mobile broadband required high throughput with high coverage capacity, many IoT-related application requirements are quite unlike the parameters concerned, for example, the requirements of health-related IoT devices are low latency and high reliability. The dynamics of such multidimensional demand requirements of IoT applications has led 5G to distinguish the applicability of two classes as massive machine-type communications (mMTCs) and ultra-reliable low-latency communications (URLLCs) [6]. Further, latency- and reliability-based application can be divided based on the individual requirements, such as factory automation, process automation, smart grids, intelligent transport systems, and professional audio. It should be considered as a part of the future IoT ecosystem when the application domain of the 5G network is defined.

The most commonly used application is enhanced mobile broadband. 5G is built with higher quality communication, which is faster and more comfortable for consumers. The new-generation technologies require smart machine-to-machine communications. Therefore, 5G provides enhancements such as speed, latency, efficient energy consumption, and communication. In a driverless car we can also implement 5G technologies for efficient communication with the surroundings. 5G enables various real-time applications such as cloud services, internet of vehicles (IoV), financial technology, industry automation, smart cities, and buildings. It has many potential applications ranging from ultrahigh definition video to virtual reality applications.

3 Novel Architectures and Implications

The term "hyper connectivity" is necessitated to serve the application, which is ready, growing fast, and should be accommodated into 5G. Applications such as automatic cars, high-resolution video streaming, augmented reality and virtual reality, and many more, have sought high throughput. At the same time, the research community is looking toward SDN and network function virtualization (NFV) is a promising field for serving the newly developed technology such as massive MIMO, millimeter wave, ultra-duplex channel and beam-forming as the basic design principles of 5G. In Agyapong et al. [7], two-layer architecture is given, which enables radio networks to be accommodated and integration with the network cloud (Table 1).

In the field network cloud to serve the enterprise and individuals with high-throughput on-demand services for the 5G network, many organizations are working to achieve comprehensive cloud adaptation. To this end, the report "A Cloud-Native 5G Architecture is Key to Enabling Diversified Service Requirements" was published by Huawei [5] in 2016.

The potential of 5G architecture is growing fast, and many proposals [8–12] have come up along the way, including applications such as the cloud, SDN controller cloud, SDN-based C-RAN, SDN-based transport network, and the SDN-based core network. Moreover, several promising technologies, which have been forwarded

Table 1 Name and abbreviation used in the text

Abbreviation	Name
ALTO	Application Layer Traffic Optimization
API	Application Programming Interface
BBU	Baseband Unit
BS	Base Station
BSS	Business Support System
CC	Cooperative Communication
CCNC	Cooperative Communication and Network Coding
CLD	Cross-Layer Design
CR	Cognitive Radio
D2D	Device-to-Device
ETSI	European Telecommunications Standards Institute
FD	Full Duplex
FDD	Frequency Division Duplex
GBP-LS	Group-Based Policy Label Switched
GMPLS	Generalized Multi-Protocol Label Switching
IETF	Internet Engineering Task Force
IoT	Internet of Things
IoV	Internet of Vehicles
ITU	International Telecommunications Union
LISP	Locator/ID Separation Protocol
MEC	Multi-access Edge Computing
MIMO	Multiple Input and Multiple Output
mMTC	massive Machine-Type Communications
MNO	Management and Network Orchestration
MP	Multipath Propagation
MPLS	Multi-Protocol Label Switching
MWC	Millimeter Wave Communications
NBI	North-Bound Interface
ND	Network Device
NFV	Network Function Virtualization
NFVRG	NFV Research Group
NOMA	Non-Orthogonal Multiple Access
NOS	Network Operating System
ODL	OpenDayLight
OFDMA	Orthogonal Frequency Division Multiple Access
ONF	Open Networking Foundation
ONOS	Open Network Operating System
OPNFV	Open Source Project of NFV
OSS	Operation Support System
PCEP	Path Computation Element Protocol
PCEPs	Path Computation Element Protocol Secured
SDN	Software Defined Network

(continued)

Table 1 (continued)

Abbreviation	Name
SFC WG	Service Function Chaining Working Group
STC	Space–Time Coding
TDD	Time Division Duplex
3GPP	3rd Generation Partnership Project Service
URLLC	Ultra-Reliable Low-Latency Communication
VNF	Virtualized Network Framework
VTN	Virtual Tenant Network
XLM	Cross-Layer Module

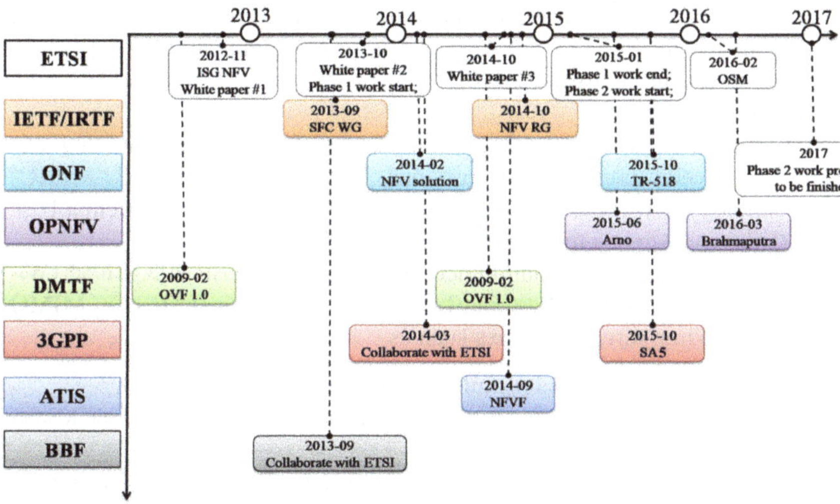

Fig. 1 Standardization time-line [15]

as the backbone for research into and the development of 5G have also been listed in these proposals [8–12]. These include NOMA [13, 14], massive MIMO [17], full duplex (FD), device-to-device (D2D) communication, millimeter wave communications (MWCs), automated network organization, cognitive radio (CR), cooperative communications (CCs) and network coding, green communication, and many more. These technologies are broad in nature; therefore, the conceptual implication of some ongoing research trends in the development of 5G architecture are explained briefly in this section. In Fig. 1, the timeline of development with the corresponding agency responsible for the individual technology is illustrated.

a. *NOMA*
   NOMA is a multiple access technique for the 5G cellular wireless network. It is capable of serving multiple user equipment using single 5G-NB (node base or base station). It helps various users at the same time/frequency resources.

The existing radio access technology, orthogonal frequency division multiple access (OFDMA), has had many upgrades, although its limitations, such as spectrum deficiency, have been major inspiration for NOMA techniques (power and code domain). It has been shown in many pieces of research that NOMA can fulfill the requirement of 5G at the network level in addition to the data rate at the user level. The advancement of NOMA can achieve desirable capacity handling, power and spectrum efficiency, interference handling, pairing, and delay with fairness. Moreover, the incorporation with wireless communication such as CCs, massive MIMO, beamforming, space–time coding (STC), and NC helps in dealing with the implementation for other communication networks (wireless networks, ad-hoc wireless networks, wireless sensor networks) apart from the cellular communication. The important issues of limitations and directions of future research with regard to NOMA were listed in Islam et al. [13, 16], which can be helpful for further exploration.

b. *Massive MIMO*

In radio, MIMO is used to multiply the capacity of a radio link that can transmit and receive multiple transmissions. Multi-user MIMO (MU-MIMO) is old technology and the adoption of the basic concept of multi-antenna base stations (BSs) is to serve a multitude of users.

MIMO accepted in wireless communication along with IEEE 802.11n/ac (Wi-Fi), HSPA+ (3G), WiMAX (4G), and Long Term Evolution (LTE 4G). MIMO can be classified into three categories such as:

- Pre-coding, which is useful for multi-stream beamforming
- Spatial multiplexing handles a high-rate signal and is split into multiple lower-rate streams
- Diversity coding is useful if there is no information available regarding the channel at the transmitter.

Sometimes there is the misconception: "Isn't massive MIMO just MU-MIMO with more antennas?"

The answer to this is "No."

The key difference between massive MIMO and conventional MU-MIMO is not only the number of antennas. These key differences are listed in Table 2

Marzetta's massive MIMO [17] concept is the way to deliver the theoretical gains of MU-MIMO under practical circumstances. To achieve this goal, we need to acquire accurate channel state information, which in general can only be done by exploiting uplink pilots and channel reciprocity in time division duplex (TDD) mode. Using channel hardening and favorable propagation phenomena, the system operation can also be simplified in massive MIMO.

The single fundamental issue in 5G is to handle the multipath propagation (MP) efficiently, and MIMO has already established technology for MP in the past. Therefore, the exploitation of MIMO technology on large-scale antennas (massive MIMO) equipped with the channel state information provides a scalable technology.

Table 2 The key differences between conventional MU-MIMO and massive MIMO

Comparison parameters	Conventional MU-MIMO	Massive MIMO
Base station (BS) antennas (M) and users (K) limitation	M ≈ K and both are small (e.g., below 10)	M ≫K and both can be large (e.g., M = 100 and K = 20)
Duplexing mode	Designed to work with both TDD and frequency division duplex (FDD) operation	Designed for TDD operation to exploit channel reciprocity
Channel acquisition	Mainly based on codebooks with a set of predefined angular beams	Based on sending uplink pilots and exploiting channel reciprocity
Link quality after precoding/combining	Varies over time and frequency, due to frequency-selective and small-scale fading	Almost no variations over time and frequency, thanks to channel hardening
Resource allocation	It must change rapidly to account for channel quality variations	It can be planned in advance because the channel quality varies slowly
Cell-edge performance	BS cooperation is required for good performance	Cell-edge signal-to-noise ratio increases proportionally to the number of antennas, without causing more inter-cell interference

The key research for 5G in the field of massive MIMO is to achieve high spectral efficiency along with coverage enhancement. In this context, an algorithm whose complexity is low and that is able to enhance in terms of channel prediction is desirable. The detailed description about the massive MIMO and technological aspect is given in Chap. 6

c. *Millimeter wave communication (MWC)*

In MWC, the band of the very high-frequency spectrum (between 30 and 300 GHz) is used for testing and development of 5G infrastructure for wireless communication as suggested by the International Telecommunications Union (ITU). The millimeter wave spectrum is no longer used owing to its limitations (short wavelengths that range from 10 to 1 mm), such as high atmospheric attenuation. At the same time, the broadness in the range also passes the feasibility test to accommodate a larger number of devices, which is a requirement of high-speed wireless communication. Moreover, this issue poses other challenges in terms of circuits and system design, interference management, spatial reuse, anti-blockage, and dynamics control. An extensive study with recent standard with architectural design guidelines is given in Niu et al. [18]. Moreover, the research highlights of MWC can be summarized as beamforming architecture, signal attenuation due to free space propagation, blockage effect, narrow beams in a non-orthogonal D2D (for booster/small cell) transmission for the heterogeneous/homogeneous network.

The limitation of short-range line of sight traveling is the main obstacle to the adoption of MWC in 5G. Although much research is carried out to tackle these problems, such as in Rappaport et al. [19], a new MWC for cellular systems has been proposed with proper hardware specifications to use a frequency range of 28- and 38-GHz with directional antennas at the base station and mobile devices. In Roh et al. [20], a hybrid beamforming scheme and its link- and system-level simulation results have shown the feasibility of MW bands for cellular communication.

d. *Cooperative communications and network coding (CCNC)* In wireless communication, storing and forwarding of the packet is one of the primary challenges. With the addition of MIMO on many fronts in 5G, the complexity is increased even further. In CCNC, it can store and forward mix packets. Moreover, it can pass other parameters (throughput, latency, energy consumption, robustness) according to the requirements of 5G with consideration of D2D communication and massive connectivity. In Vieira and Vieira [21], CCNC advantages, and applications have been covered in the context of 5G infrastructure. The common application of CCNC can unicast, multicast, broadcast, and relay, and is extensively included in a great deal of research. Still, though, the packet loss in the dynamic spreading spectrum, using multiple channels and multiple antennas is one of the significant research challenges that needs future attention.

4 Cross-Layer Design

Traditionally, all the layers are encapsulated in an individual module at the protocol stack. The encapsulation, at the same, prevents the free flow of information within the module and creates hurdles in many applications. The interactions between different non-adjacent layers allow the communication architecture to work as a system instead of the different protocol stacks and it is termed cross-layer design (CLD).

Very few proposals have been available on prospective CLDs for the 5G network, and some of the proposals have tried in the past [22–25] to explore the possibilities of CLDs in 5G, although, several texts have provided the perspective of CLD progress in the past and open research issues for the future in 5G, wireless networks [26, 27], and wireless sensor networks [28–31]. In Mendes and Rodrigues [28], a vast survey on CL solution for wireless networks has been provided by maintaining taxonomy based on technology used at different layers. The key element of CLD is very broad in nature for any type of communication paradigm; therefore, in this text our approach will be to discuss the proposal made for CLDs in 5G, instead of covering elements of CLD itself.

The significance of information sharing among network stack or layers is necessary for performance improvement. With this fact, in Shiab [22], a CLD proposal has been made for SDN and has gained a great deal of momentum because of the flexibility it offers by creating re-programmable networks that adapt quickly and efficiently to the changing demands of today's networks.

In SDN, decoupling of the control and data planes is already being accepted in the research fraternity; therefore, a survey and suggestion have been made to design a logically centralized controller that has a global overview of the whole network that can achieve and assign the control task. Motivated by the great benefits of both paradigms, CL SDN architectures solve challenges in different applications.

In Niu et al. [23], for heterogeneous networks, a software-defined mmWave mobile broadband system via a CLD is described, in which it deploys small cells in the mmWave band underlying the macrocell network. Different than the communication systems using lower carrier frequencies, MWCs have unique features, such as high propagation loss, directional communications, and sensitivity to blockage. It also overcomes the challenging problems in mmWave networks, such as interference management, spatial reuse, anti-blockage, QoS guarantee, and load balancing. In this architecture, a centralized controller is introduced by abstracting the control functions from the network layer to the physical layer. A quantitative simulation in a realistic indoor scenario has been demonstrated for a better understanding of the performance advantages of the CLD SDN in terms of network throughput and flow throughput. It is one of the first cross-layer and software-defined designs for MWCs, which opens up the opportunity for MWCs to make a significant impact on future 5G networks.

In Baligh et al. [24], CLD provisions for the cellular network are given to accomplish virtualized and mass-scale cloud architectures. In this 5G-based network, all the nodes are connected via a backhaul network and managed centrally by such cloud centers. The significant computing power made available by the cloud technologies has enabled the implementation of sophisticated signal processing algorithms, especially by way of parallel processing, for both interference management and network provision. The major signal processing tasks for 5G due to the increased level of frequency sharing, node density, interference, and network congestion are covered in the study.

In Tang et al. [25], CL resource allocation with elastic service scaling in a cloud radio access network is described. It is aimed at improving the spectrum and energy efficiency of wireless networks by migrating conventionally distributed base station functionalities into a centralized cloud baseband unit (BBU) pool. In a cross-layer resource allocation model for CRAN, the minimization of the overall system power consumption in the BBU pool, fiber links, and the remote radio heads are considered an objective function such as mixed-integer nonlinear programming.

The heterogeneity of the IoT devices can easily be handled by CLDs. The main principle of CLDs is to provide a better performance metric and serve the requirement of the applications. In Sah and Amgoth [32], CLD aspects of wireless sensor networks are presented. Many sensors will be part of IoT applications whose performance requirement will be different, as explained in the previous section. Therefore, proposals such as XLP, XLM, DGRAM, which are of modular design by nature, can serve as the virtual backbone in the sub-context of IoT networks.

In Ranjan and Varma [29, 30], an architectural view of CLDs is given with the issues involved in implementation. The classification based on the design module is given as conventional, complex, and unified. The working module gives an example of understanding the significance of CLDs. One of the important contributions of CLDs for wireless sensor networks is provided [33], which gives a cross-layer module (XLM). In this, traffic, congestion, and errors are handled carefully.

5 SDN-NFV-Based Models

The wireless and telecommunication infrastructure has been well established in past decades, although the load on the network is still growing exponentially. The estimated cost of infrastructure is one trillion dollars, which poses a challenge in the adoption of any newly emerged technology. To overcome this significant cost, adoption of the hardware seems infeasible and searches for software-defined infrastructure has become the impetus for now or never transformation.

There is dissent in the categorization of software-defined networks (SDNs), and network function virtualization (NFV) as some suggestions have been made and proceeds by considering these two as separate entities. Although, some definition is given with NFV as a subset of SDN. A great deal of research [34–46] work, along with institute and company collaboration, is going to smooth an insusceptible transformation. In this section, we discuss the general key elements in the context of the SDN-NFV model. Further, some of the research proposals with their contributions are briefly discussed to enable understanding of the recent advancement.

Although SDN and NFV are separate initiatives, it appears that they are complementary to each other. Our approach is to deal with these two entities separately and highlights the connectivity among them wherever it is needed.

5.1 Software-Defined Network (SDN)

The term SDN was initially coined to represent the ideas and work around OpenFlow at Stanford University, Stanford, CA, USA [40]. The intention of the original defined SDN model was to provide the network architecture in which the data plane can be managed remotely to enhance the flexibility, although many times the network industry has shifted perspective according to the requirements and anything that involves the software. Today, in the new age of the 5G framework where it is advancing to accommodate a large number of devices for a different application, the SDN is viewed as a promising adoption.

According to the suggestion made in Newman et al., Gude et al., Jamjoom et al., McKeown et al., and Greene [40–43, 47], the SDN can be defined based on the following key attributes:

- Decoupling of control and data planes to ensure the reduction in the control to the network device (ND) and conclude the task of ND to a simple data forwarding element.
- Unification for different network devices such as routers, switches, firewalls, and middleboxes to ensure the flow-based forwarding. These abstractions enable the same services for similar flow, and differentiation based on the destination will disable them. Adoption of flow-based forwarding helps with the enhancement of flexibility.
- The abstraction of the control entity moved to an SDN controller that handles resource allocation in a centralized manner.
- The network is programmable through software applications running on the top SDN controller, which is the only responsible entity to interact with data plane devices.

In Rotsos et al. [44], the SDN architecture with its basic building blocks is described, which is generalized based on the Open Networking Foundation (ONF) [45] for a new form of network. In Blanco et al. [39] technology pillars in the architecture of future 5G mobile networks are described, NFV, multi-access edge computing (MEC), and SDN, which has been mostly inspired by the road map given in "IMT Vision" in 2015 [48]. In Kreutz et al. [46], a simplified view of an SDN architecture, a layered view of networking functionality and its fundamental abstractions are explained in detail, and we refer the interested reader to this survey. Meanwhile, for a simplistic explanation to cover the basics of SDN architecture, we take Rotsos et al. [44] into account. In Fig. 2, an architectural model of an SDN control stack is illustrated in the form of a block diagram. The basic building blocks of SDN architecture consist of three components as follows:

a. Data plane
b. Control plane
c. Application plane

The data plane accommodates all network devices such as the gateway, router, etc. The traffic monitoring and packet manipulation functions are limited to their lowest levels to ensure the restricted functionality. It is equipped with the interface to communicate with and receive instructions from the control plane. The suggestion for the categorization of the control interface also varies with the different proposals, such as OpenFlow [49] and PCE [50], designed to manipulate the device-forwarding policy, and management interfaces, such as NETCONF [51] and OF-CONFIG [52], designed to provide remote device configuration, monitoring, and fault management.

The authoritative part of the architecture is a control plane that contains a network operating system (NOS). It comprises several functionalities, such as centralizing control of multiple data plane devices along with high-level interfaces for management applications. Moreover, topology monitoring and resource virtualization services have also been assessed by the NOS. The policy analysis regarding application, mitigation, and administration to ensure the communication conflict are

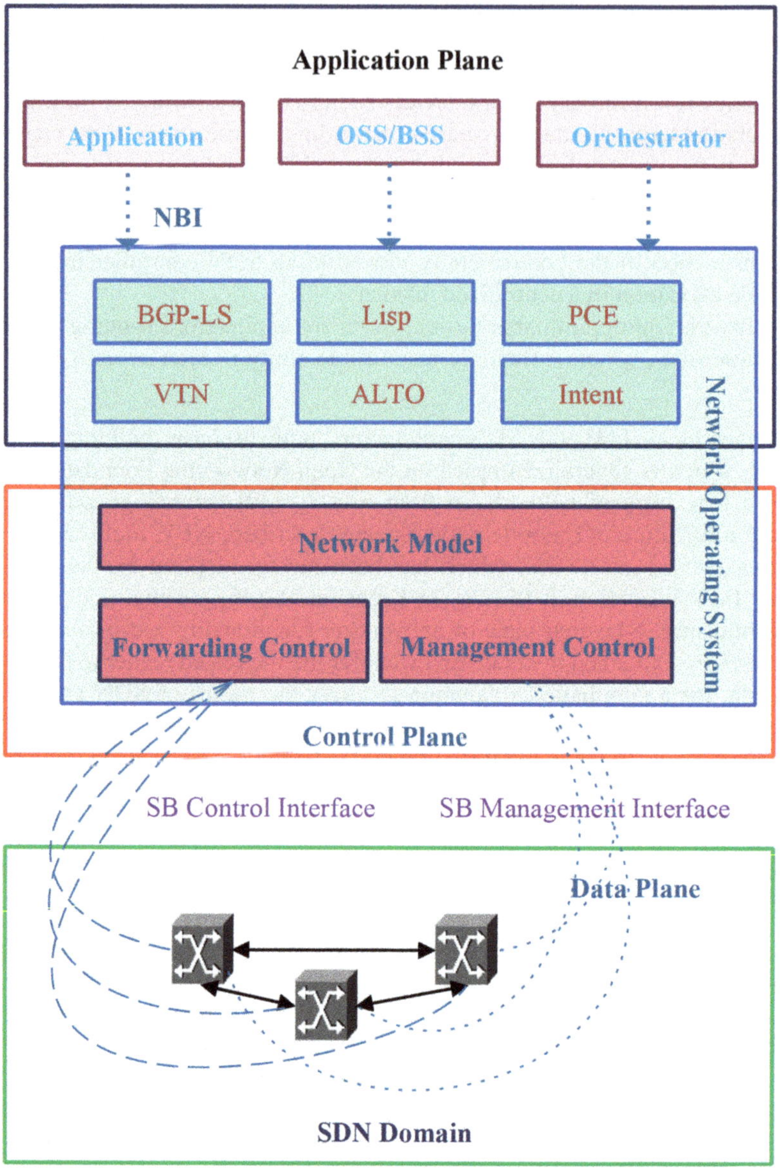

Fig. 2 The SDN architecture model can be separated into three layers: the data, control, and application planes [44]

other responsibilities of the NOS. The main advantage of the logical separation of network control from the control plane is that it allows it to encapsulate multiple individual instances of the NOS. Each of these encapsulates instance control load management and is able to handle overlapping efficiently. At the same time, multiple instances of the NOS allow SDN to have better failure management and recovery (distribution of load among other instances). For a detailed presentation of the standardization, research, and implementation efforts in SDN, the reader can refer to Kreutz et al. [46].

In the application plane, the north-bound interface (NBI) standardization is one of the major research domains. For a general understanding, we will try to cover the key elements of NBI standardization, its working and application in SDN. It is one of the crucial parts of service orchestration because in the end it is responsible for the control and monitoring of service connectivity, network resource utilization, and flexible fault management. It can be divided into two major categories. The first category contains low-level information whereas the second category includes high-level and innovative control abstractions, which are responsible for creating the interfaces based on match and forward. In low-level information management, several centralized [53, 54], distributed [55], and common information models (CoreModel) [56] suggestion made for specifications. In the high-level abstraction, the general specifications are the NOS management application, operation support system (OSS), service orchestrator, and control applications. The interface of application and control is also enabled by these entities, as can be seen in Fig. 2. Moreover, other suggestions for the specification of legacy control interfaces such as the open network operating system (ONOS) [57], OpenDayLight (ODL) [58], are also popular owing to the nature of their open source NOS.

Other elements of the application plane and its working can be summarized as follows based on the projects:

a. *Path computation element (PCE)*
   In label switching, the control technology that handles address resource and control forwarding have been specified in PCE. For distributed path establishment, protocols such as generalized multi-protocol label switching (GMPLS) and multi-protocol label switching (MPLS) have been used in PCE. Switches use different information for path establishment such as the traffic pattern of routing data in OSPF-TE [59], and signaling protocols, such as resource reservation protocol-traffic engineering in Awduche et al. [60]. These methods gather information about network resources and topology for further path establishment. At the same time, MPLS and GMPLS have limitations with regard to computation resources.
b. *Locator/ID separation protocol (LISP)*
   It [61] is a network architecture that can manage the scalability issues of routing on a large-scale. It is a dual addressing mechanism that decouples the location of a host from its unique identifier. The host equipped with LISP only requires the end-point identifier for packet communication, while routing nodes use distributed mapping services for translation of the forwarding address. The ONF is currently developing a proof-of-concept implementation of the application programming interface (API) as part of the ASPEN project [62] to provide robust LISP.
c. *Group-based policy label switched (GBP-LS)*

The GBP-LS module is an ODL project to implement and support the protocols, as mentioned above, as a control application. Moreover, it is also being suggested for use in PCE extensions, such as Stateful-PCE [63] (time, sequence, and resource usage synchronization), path computation element protocol (PCEP) for segment routing (dynamic LSP management), and secure transport for PCEP (PCEPSs) [64] (security extension).

d. *Intent-based networking*
   Intent-based networking is a popular SDN NBI exploring the applicability of declarative policy languages in network management. In this, suggestions can be made to the NOS about acceptable network states and leave low-level network configuration along with the creation of new NOS instances to the NOS. The Internet Engineering Task Force (IETF) has adopted the NEMO specifications [65], an intent-based networking policy language.
e. *Application layer traffic optimization (ALTO)*
   The ALTO [66] is an IETF-WG specification for providing end-user applications to access accurate network performance information. The vision of development is that, in the distributed application environment, major intelligence improvement is possible based on the network routing information to serve the end users. Improvement toward bandwidth-intensive or latency-sensitive applications is expected from ALTO. Its secondary work is to enable a service orchestrator to manage and action deployment decisions.
f. *Virtual tenant networks (VTNs)*
   In NEC [67], the proposal is made by NEC regarding VTNs with the vision that they will able to provide an abstraction of the network that virtually separates the virtual overlay networks from the topology of the underlying network infrastructure. It allows users to map requested topology over the physical topology and also to enable service deployment for the service orchestrator.

In Trivisonno et al. [34], the SDN-based plastic architecture for 5G mobile networks is covered, together with its functionality and compatibility. The architecture presented in Trivisonno et al. [34] describes dynamic control plane instantiation, device attachment, backward compatibility, and service request and mobility management in detail. More often, the flexibility required to support heterogeneous and homogeneous networks is one of the major expectations of the SDN-based model. Further, these requirements often produce a set of challenging issues, such as latency, dependability, reliability, and scalability. Suggestions have been made regarding the decoupling of control and data planes, which also comes with its set of gains and losses. With decoupling, network equipment is only responsible for forwarding a data packet based on the control message received and enhancing the flexibility of the architecture. Although suggestions for an SDN model are often made regarding decoupling the control and data planes, NFV, on the other hand, anticipates the instantiation of network functions for resources (hardware, software).

5.2 *Network function virtualization (NFV)*

The high level NFV framework can be generalized in the four components as follows:

a. NFV infrastructure
b. Virtualized network framework
c. OSS\business support systems (BSS) layer
d. Management and network orchestration (MNO)

NFV Infrastructure

The NFV infrastructure is a type of cloud data center containing hardware and virtual resources to build the NFV environment. It includes sensors, switches, virtual machine, and virtual switches (Fig. 3).

- *Hardware resources* It contains computing resources such as RAM and the processing unit, storage resources such as disk storage, and network resources such as a switch, firewall and routers, and many more.
- *Virtualization layer* It abstracts hardware resources and decouples the software from the hardware. Moreover, it enables the software to progress independently from the hardware. There are multiple open source and proprietary options available, such as KVM, QEMU, VMWare, etc.
- *Virtualization resources* It contains a virtual computer, virtual storage, and a virtual network.

Virtualized Network Framework (VNF)

The VNF is the basic building block of NFV architecture and its acts as a software implementation of the network function. It can connect or be combined as building blocks to offer a full-scale network communication service. When more than one VNF is connected in a string fashion to accomplish the larger task, then it is known as service chaining. Examples of VNFs are vIMS, vFirewall, and VRouter.

Management and Network Orchestration (MNO)

- *Virtual infrastructure manager* It controls and manages interaction of VNF with NFV computing, storage, and network resources. Further, it also ensures the workability of deployment and monitoring tools for virtualization layers.
- *VNF manager* It manages the life-cycle of VNF instances and is also responsible for initialization, updating, scaling, and termination of the VNF instances.
- *Orchestrator* The main task is to manage the life-cycle of network services, which includes instantiation, policy management, performance measurement, and monitoring of key performance indicators.

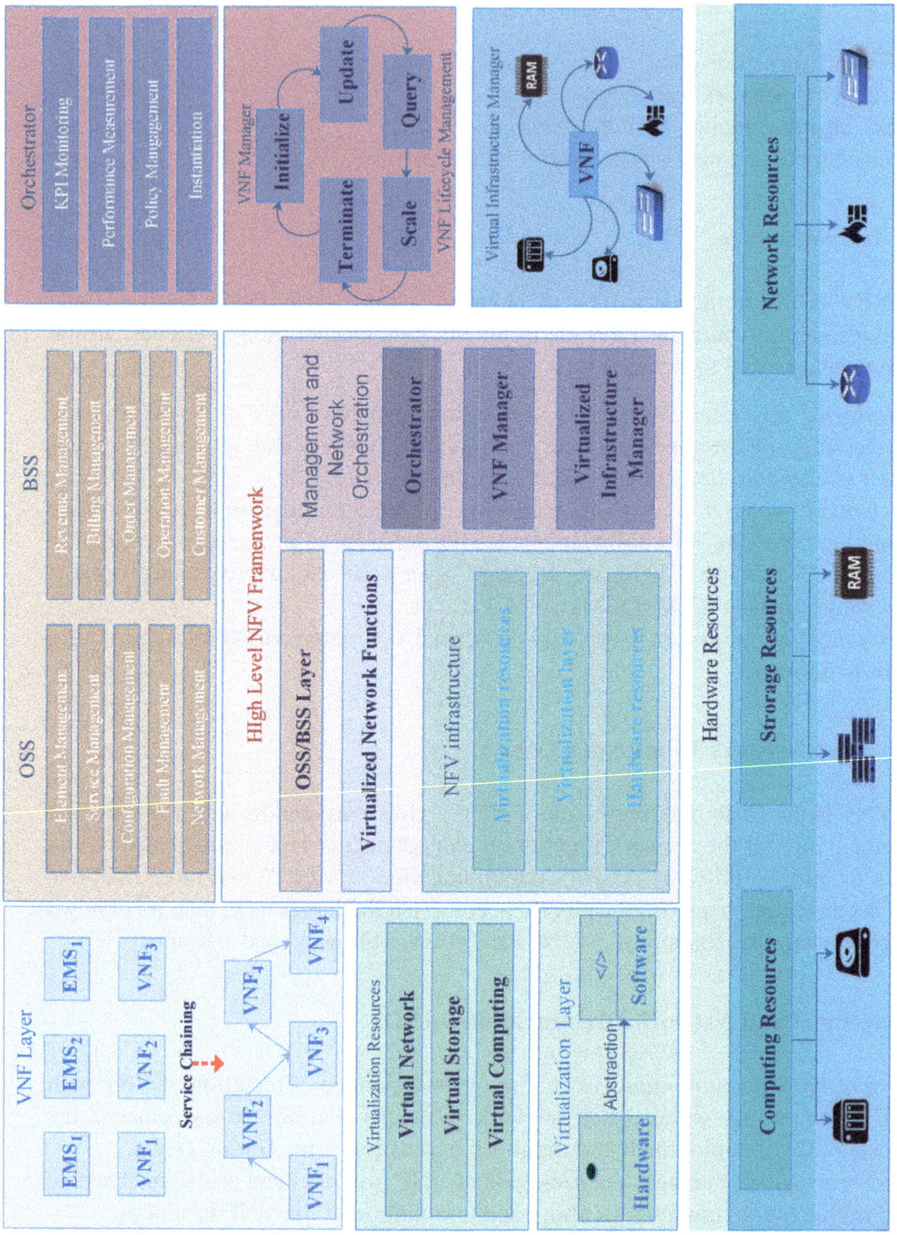

Fig. 3 The SDN architecture model can be separated into three layers: the data, control, and application planes

OSS\BSS Layer

While the OSS deals with network management fault management, configurational management, service management, and element management, the BSS administers the client's side issues, such as customer management, order management, and billing and revenue management.

6 Service Architectures and Potential Direction

At the center of most present-day systems and services is commonly a cloud- and virtualization-based stage. This is also the case for 5G systems. These are programmable stages that enable the wide range of capacities and need to be manufactured, designed, associated, and deployed on a large scale that is required at the given time. One case of such a stage is the Open Source Project of NFV (OPNFV) [68], an open source venture supporting NFV.

In any case, interest in smooth scalable frameworks that can be custom-fitted for new circumstances does not stop at the platform. The capacity to grow new capacities effortlessly, time-to-market, and the utilization of off-the-rack innovation additionally drive changes in the system capacities themselves. The objective is to move from telecom-style convention interfaces to electronic APIs. The 5G center system will be founded on what is called "Administration Based Architecture" (ABA), revolving around administrations that can enroll themselves and buy in to different services. This empowers a more adaptable improvement of new administrations, as it winds up conceivable to associate with different parts without presenting individual new interfaces. The new framework design is indicated in the 3rd Generation Partnership Project Service (3GPP) technical specification [69].

One observation that we would make is that when building cloud-native, optimized executions, it is valuable to contemplate the design of the architecture to finish everything. Present-day stages offer numerous services and conceivable outcomes that were unrealistic before, including creating process examples at locations when they are required, at a low cost and short delay. Completely advanced outlines can profit from these highlights, and we expect that the most ideal network system frameworks continue to evolve along this path.

6.1 Industry Initiatives

To begin with, some organizations in their individual capacity or with the support of regional players explore the key features of 5G, such as architectural design, technology, hardware resources, transition feasibility research, etc. This initiative overview of the standardization road map related to 5G has been summarized [39, 70] in Table 3.

Table 3 Initiative overview for the standardization road map of 5G

Initiative	Standardization description
International Telecommunications Union (ITU)	Focus on the developing the framework for future 5G systems, proposed as IMT-2020
3rd Generation Partnership Project (3GPP)	Responsible for addressing a subset of requirements that are important for current commercial needs with the features, use cases, detailed requirements, etc.
European Telecommunication Standards Institute (ETSI)	Responsible for addressing industry specifications applicable in fields such as SDN, NFV, and autonomic network management with open and inter-operable NFV ecosystem
Evolution and Ecosystem Work Group (EVE WG)	Development of the framework for feasibility studies and requirements in relation to new NFV use cases and associated technical features, interfaces, and architectures with the support of interoperability
The Institute of Electrical and Electronics Engineers (IEEE)	The initiation of the standardization work on future 5G systems. It has also started its pre-standardization research group on cloud-based mobile core to analyzing SDN/NFV concepts applied to 5G
Open Networking Foundation (ONF)	The user-driven organization that is dedicated to the promotion and adoption of SDN through the development of open standards. It has been an open, collaborative development process from the end-user perspective. Moreover, mobile networks, opportunities identification in 5G (e.g., NGMN, 3GPP) are related to SDN. The main outcome of ONF is the OpenFlow standard, which is one of the first SDN standards. It enables remote programming of the forwarding plane from a centralized control plane element. ONF encourages conformity certification for devices supporting OpenFlow. A number of hardware vendors such as Cisco, Juniper, Dell, HP, BigSwitch Networks, Brocade Communications, and Alcatel-Lucent include OpenFlow implementation in some of their products
FP-7 (Europe)	The Radio Access and Spectrum (RAS) is a cluster activity that comprises the research efforts on radio access and spectrum in the area of future networks in the EU. This cluster is analyzing architectural aspects of 5G mobile and wireless communication systems, considering both wireless and wired parts, targeting a fully integrated solution
Public Private Partnership Program:	Many individual countries (5G-PPP (Europe), 5G America's, 5GMF (Japan), 5G Promotion Group (China), 5G Forum (Korea)) have taken the initiative in development and ease of adoption for transformation. The objective is to participate in the current architectural design process, SDN/NFV adoption, mobile communications infrastructure, regulatory considerations, etc.
NGMN:	It is responsible for the process needed for mobile network operators to establish clear functionality and performance targets in addition to fundamental requirements for deployment scenarios and network operations, and leading to the implementation of a cost effective network evolution

(continued)

Table 3 (continued)

Initiative	Standardization description
Small Cell Forum (SCF)	It is aimed at driving the wide-scale adoption of small cells and to influence and deliver technical inputs that inform and enhance the standards process. The priorities of SCF include the understanding of future network transformations with a particular focus on virtualization of the small cell layer and the preparation of small cell technology for mass deployment in heterogeneous networks exploiting self-organizing capabilities
Network Function Virtualization Research Group (NFVRG)	It explores new directions in the development and collaboration of NFV deployments and applications. This group takes into account the benefits of the SDN paradigm applied to NFV
Software Defined Networking Research Group (SDNRG)	Works in the definition of models, interfaces, abstractions, metrics, operation of network devices, protocols, among others, to leverage the SDN concept
I2RS	The Working Group I2RS defines use cases and a basic architecture based on blocks to increase the routing system (hardware and software) based on a set of interfaces and protocols. I2RS takes into account the SDN and NFV requirements

In this section, the potential directions are listed based on the suggestions made by ongoing research and development.

a. Interference management and cancellation in 5G networks in an FD radio network is still an open problem for multi-cell.
b. Hand-off management in 5G networks.
c. QoS management in 5G networks as it is supposed to satisfy the highest level of QoS.
d. Load balancing in 5G networks as channel access control management is still an open research issue.
e. Security and privacy in 5G networks as current security and privacy solutions are unable to handle massive connections.

7 Conclusion

In this chapter, the requirement for a transition from existing networks to 5G and its key technological progression is discussed. In most of the 5G-based architecture, SDN and NFV have been gradually used to implement network hardware virtualization. The primary consideration is to retain a conventional operational model and make a smooth transition in software architecture. 5G networks require continuous innovation through cloud adoption to customize network functions. In this chapter, a detailed insight and ongoing research into CLDs, SDN-based models, and NFV models are explained. Finally, service architecture and potential research into the improvement of 5G technology have been suggested.

Appendix

NEMO

It is a domain-specific language (DSL), following the declarative programming paradigm. Its progressing project is to make abstract specification on an end-point, describe a network end-point, a connection, describe connectivity requirements between network end-points, and an operation describes packet operations.

Huawei is currently leading an implementation initiative, based on ODL and the OPNFV project [71]. In parallel, the ONF has recently organized a work group to standardize a common intent model. The group aims to fulfill two objectives:

a. Define the architecture and requirements of intent implementations across controllers and define portable intent expressions.
b. Develop a community-approved information model that unifies intent interfaces across controllers.

The particular standard is combined with the improvement of the Boulder structure [72], an open-source [73], OpenStack Neutron [74], and portable intent system that can be incorporated into all major SDNs. Its intent toward the goals through a grammar, which comprises subjects, predicates, and targets. The dialect can be extended to incorporate imperatives and conditions. The reference support execution has set up similarity with ODL through the Network Intent Composition (NIC) project, whereas ONOS support is currently being worked on.

References

1. J.C. Lin, Synchronization requirements for 5G: an overview of standards or specifications for cellular networks. IEEE Veh. Technol. Mag. **13**, 1–1 (2018)
2. S.K. Biswash, D.N.K. Jayakody, Performance based user-centric dynamic mode switching and mobility management scheme for 5G networks. J. Netw. Comput. Appl. **116**, 24–34 (2018)
3. S.A. Hassan, M.S. Omar, M.A. Imran, J. Qadir, D.N.K. Jayako, Universal access in 5G networks, potential challenges and opportunities for urban and rural environments, in *5G Networks: Fundamental Requirements, Enabling Technologies, and Operations Management*, ed. by A. Al-Dulaimi, X. Wang, I. Chih-Lin (Wiley, Hoboken, 2018)
4. M. Agiwal, A. Roy, N. Saxena, Next generation 5G wireless networks: a comprehensive survey. IEEE Commun. Surv. Tutorials **18**(3), 1617–1655 thirdquarter (2016)
5. Huawei, 5G network architecture – a high-level perspective. (2016) http://www.huawei.com/en/industry-insights/mbb-2020/trends-insights/5g-network-architecture
6. P. Schulz, M. Matthe, H. Klessig, M. Simsek, G. Fettweis, J. Ansari, S.A. Ashraf, B. Almeroth, J. Voigt, I. Riedel et al., Latency critical IoT applications in 5G: perspective on the design of radio interface and network architecture. IEEE Commun. Mag. **55**(2), 70–78 (2017)
7. P.K. Agyapong, M. Iwamura, D. Staehle, W. Kiess, A. Benjebbour, Design considerations for a 5G network architecture. IEEE Commun. Mag. **52**(11), 65–75 (2014)
8. S.B.H. Said, M.R. Sama, K. Guillouard, L. Suciu, G. Simon, X. Lagrange, J.-M. Bonnin, New control plane in 3GPP LTE/EPC architecture for on-demand connectivity service, in *2013 IEEE 2nd International Conference on Cloud Networking (CloudNet)* (IEEE, 2013), pp. 205–209

9. Z. Wang, W. Zhang, A separation architecture for achieving energy-efficient cellular networking. IEEE Trans. Wirel. Commun. **13**(6), 3113–3123 (2014)
10. C.J. Bernardos, A.D.L. Oliva, P. Serrano, A. Banchs, L.M. Contreras, H. Jin, J.C. Zúñiga, An architecture for software defined wireless networking. IEEE Wirel. Commun. **21**(3), 52–61 (2014)
11. J. Costa-Requena, SDN integration in LTE mobile backhaul networks, in *2014 International Conference on Information Networking (ICOIN)* (IEEE, 2014), pp. 264–269
12. Z. Ma, Z. Zhang, Z. Ding, P. Fan, H. Li, Key techniques for 5G wireless communications: network architecture, physical layer, and MAC layer perspectives. Sci. China Inf. Sci. **58**(4), 1–20 (2015)
13. S.M.R. Islam, N. Avazov, O.A. Dobre, K.-S. Kwak, Power-domain non-orthogonal multiple access (NOMA) in 5G systems: potentials and challenges. IEEE Commun. Surv. Tutorials **19**(2), 721–742 (2017)
14. S. Qureshi, S.A. Hassan, D.N.K. Jayakody, Divide-and-allocate: an uplink successive bandwidth division NOMA system. Trans. Emerg. Telecommun. Technol. **29**(1), e3216 (2017)
15. B. Yi, X. Wang, K. Li, M. Huang et al., A comprehensive survey of network function virtualization. Comput. Netw. **133**, 212–262 2018
16. S.M. Islam, M. Zeng, O.A. Dobre, NOMA in 5G systems: exciting possibilities for enhancing spectral efficiency (2017). arXiv preprint:1706.08215
17. T.L. Marzetta, Massive MIMO: an introduction. Bell Labs Tech. J. **20**, 11–22 (2015)
18. Y. Niu, Y. Li, D. Jin, L. Su, A.V. Vasilakos, A survey of millimeter wave communications (mmwave) for 5G: opportunities and challenges. Wirel. Netw. **21**(8), 2657–2676 (2015)
19. T.S. Rappaport, S. Sun, R. Mayzus, H. Zhao, Y. Azar, K. Wang, G.N. Wong, J.K. Schulz, M. Samimi, F. Gutierrez Jr., Millimeter wave mobile communications for 5G cellular: it will work! IEEE Access **1**(1), 335–349 (2013)
20. W. Roh, J.-Y. Seol, J. Park, B. Lee, J. Lee, Y. Kim, J. Cho, K. Cheun, F. Aryanfar, Millimeter-wave beamforming as an enabling technology for 5G cellular communications: theoretical feasibility and prototype results. IEEE Commun. Mag. **52**(2), 106–113 (2014)
21. L.F.M. Vieira, M.A.M. Vieira, Network coding for 5G network and D2D communication, in *Proceedings of the 13th ACM Symposium on QoS and Security for Wireless and Mobile Networks* (ACM, 2017), pp. 113–120
22. I. Al Shiab, Cross-layer software defined networks: a survey.
23. Y. Niu, Y. Li, M. Chen, D. Jin, S. Chen, A cross-layer design for a software-defined millimeter-wave mobile broadband system. IEEE Commun. Mag. **54**(2), 124–130 (2016)
24. H. Baligh, M. Hong, W.-C. Liao, Z.-Q. Luo, M. Razaviyayn, M. Sanjabi, R. Sun, Cross layer provision of future cellular networks (2014). arXiv preprint: 1407.1424
25. J. Tang, W.P. Tay, T.Q.S. Quek, Cross-layer resource allocation with elastic service scaling in cloud radio access network. IEEE Trans. Wirel. Commun. **14**(9):5068–5081 (2015)
26. B. Fu, Y. Xiao, H. Deng, H. Zeng, A survey of cross-layer designs in wireless networks. IEEE Commun. Surv. Tutorials **16**(1), 110–126 (2014)
27. X. Lin, N.B. Shroff, R. Srikant, A tutorial on cross-layer optimization in wireless networks. IEEE J. Sel. Areas Commun. **24**(8), 1452–1463 (2006)
28. L.D.P. Mendes, J.J.P.C. Rodrigues, A survey on cross-layer solutions for wireless sensor networks. J. Netw. Comput. Appl. **34**(2), 523–534 (2011)
29. R. Ranjan, S. Varma, Challenges and implementation on cross layer design for wireless sensor networks. Wirel. Pers. Commun. **86**(2), 1037–1060 (2016)
30. I. Al-Anbagi, M. Erol-Kantarci, H.T. Mouftah, A survey on cross-layer quality-of-service approaches in WSNS for delay and reliability-aware applications. IEEE Commun. Surv. Tutorials **18**(1), 525–552 (2016)
31. R. Muraleedharan, L.A. Osadciw, Security: cross layer protocol in wireless sensor network, in *INFOCOM 2006. 25th IEEE International Conference on Computer Communications. Proceedings* (IEEE, 2006), pp. 1–2
32. D.K. Sah, T. Amgoth, Parametric survey on cross-layer designs for wireless sensor networks. Comput. Sci. Rev. **27**, 112–134 (2018)

33. M.C. Vuran, I.F. Akyildiz, XLP: a cross-layer protocol for efficient communication in wireless sensor networks. IEEE Trans. Mob. Comput. **9**(11), 1578–1591 (2010)
34. R, Trivisonno, R, Guerzoni, I, Vaishnavi, D. Soldani, SDN-based 5G mobile networks: architecture, functions, procedures and backward compatibility. Trans. Emerg. Telecommun. Technol. **26**(1), 82–92 (2015)
35. Network Functions Virtualisation, SDN and openflow world congress (2012)
36. ETSI, Network function virtualisation-white paper2 (2013). http://portal.etsi.org/NFV/NFV_White_Paper2.pdf
37. NFVISG ETSI, Network functions virtualization, white paper (2014). http://www.esti.org/technologiescluster/technologies/nfv
38. P. Demestichas, A. Georgakopoulos, D. Karvounas, K. Tsagkaris, V. Stavroulaki, J. Lu, C. Xiong, J. Yao, 5G on the horizon: key challenges for the radio-access network. IEEE Veh. Technol. Mag. **8**(3), 47–53 (2013)
39. B. Blanco, J.O. Fajardo, I. Giannoulakis, E. Kafetzakis, S. Peng, J. Pérez-Romero, I. Trajkovska, P.S. Khodashenas, L. Goratti, M. Paolino et al., Technology pillars in the architecture of future 5G mobile networks: NFV, MEC and SDN. Comput. Stand. Interfaces **54**, 216–228 (2017)
40. K. Greene, TR10: software-defined networking. Technology Review (MIT) (2009)
41. P. Newman, G. Minshall, T.L. Lyon, IP switching-ATM under IP. IEEE/ACM Trans. Networking (TON) **6**(2), 117–129 (1998)
42. N. Gude, T. Koponen, J. Pettit, B. Pfaff, M. Casado, N. McKeown, S. Shenker, NOX: towards an operating system for networks. ACM SIGCOMM Comput. Commun. Rev. **38**(3), 105–110 (2008)
43. N. McKeown, T. Anderson, H. Balakrishnan, G. Parulkar, L. Peterson, J. Rexford, S. Shenker, J. Turner, Openflow: enabling innovation in campus networks. ACM SIGCOMM Comput. Commun. Rev. **38**(2), 69–74 (2008)
44. C. Rotsos, D. King, A. Farshad, J. Bird, L. Fawcett, N. Georgalas, M. Gunkel, K. Shiomoto, A. Wang, A. Mauthe et al., Network service orchestration standardization: a technology survey. Comput. Stand. Interfaces **54**, 203–215 (2017)
45. Open Networking Foundation, Software-defined networking: the new norm for networks. ONF White Pap. **2**, 2–6 (2012)
46. D. Kreutz, F.M.V. Ramos, P.E. Verissimo, C.E. Rothenberg, S. Azodolmolky, S. Uhlig, Software-defined networking: a comprehensive survey. Procee. IEEE **103**(1), 14–76 (2015)
47. H. Jamjoom, D. Williams, U. Sharma, Don't call them middleboxes, call them middlepipes, in *Proceedings of the third workshop on Hot topics in software defined networking* (ACM, 2014), pp. 19–24
48. M. Series, IMT vision–framework and overall objectives of the future development of IMT for 2020 and beyond (2015)
49. Open Network Foundation, Openflow switch specifications 1.5.0. https://www.opennetworking.org/images/stories/downloads/sdn-resources/onf-specifications/openflow/openflow-switch-v1.5.1.pdf
50. J.P. Vasseur, J.L. Le Roux, Path computation element (PCE) communication protocol (PCEP). Technical report (2009)
51. R. Enns, M. Bjorklund, J. Schoenwaelder, Network configuration protocol (NETCONF). Network (2011). http://www.rfc-editor.org/info/rfc6241
52. Open Network Foundation, Of-config 1.2: openflow management and configuration protocol (2014). https://www.opennetworking.org/images/stories/downloads/sdn-resources/onf-specifications/openflow-config/of-config-1.2.pdf
53. ITU, ITU-T recommendation M.3100: generic network information model. ITU **1**, 1–6 (2005)
54. ITU, M.3102: unified generic management information model for connection-oriented and connectionless networks. ITU-T **1**, 1–6 (2011)
55. DMTF Common Information Model, DMTF **1**, 1–6. http://www.dmtf.org/standards/cim

56. Open Network Foundation, Core information model (coremodel). https://www.opennetworking.org/images/stories/downloads/sdn-resources/technicalreports/ONF-CIM_Core_Model_base_document_1.1.pdf
57. P. Berde, M. Gerola, J. Hart, Y. Higuchi, M. Kobayashi, T. Koide, B. Lantz, B. O'Connor, P. Radoslavov, W. Snow et al., Onos: towards an open, distributed SDN OS, in *Proceedings of the third workshop on Hot topics in software defined networking* (ACM, 2014), pp. 1–6
58. J. Medved, R. Varga, A. Tkacik, K. Gray, OpenDaylight: towards a model-driven SDN controller architecture, in *2014 IEEE 15th International Symposium on a World of Wireless, Mobile and Multimedia Networks (WoWMoM)* (IEEE, 2014), pp. 1–6
59. D. Katz, K. Kompella, D. Yeung, Traffic engineering (TE) extensions to OSPF version 2. Technical report (2003)
60. D. Awduche, L. Berger, D. Gan, T. Li, V. Srinivasan, G. Swallow, RSVP-TE: extensions to RSVP for LSP tunnels. Technical report (2001)
61. D. Farinacci, V. Fuller, D. Meyer, D. Lewis, The locator/ID separation protocol (LISP). Technical report (2013)
62. OTF, Project ASPEN: real time media interface specification.
63. E. Crabbe, R. Varga, J. Medved, I. Minei, PCEP extensions for stateful PCE, internet-draft draft-ietf-pce-stateful-pce-14. Internet Eng. Task Force **1**, 1–6 (2017)
64. Q. Wu, D. Dhody, D. Lopez, O.G. de Dios, Secure transport for PCEP, internet-draft draft-ietf-pce-pceps-10. Internet Eng. Task Force **1**, 1–6 (2018)
65. S. Hares, Intent-based nemo overview. Internet Draft RFC **1**, 1–6 (2015)
66. E.W. Burger, J. Seedorf, Application-layer traffic optimization (ALTO) problem statement (2009)
67. NEC NEC, Programmableflow: redefining cloud network virtualization with openflow. https://www.necam.com/whitepapers/Docs/?S=Pflow
68. E.T. Docket, Technical report, Technical Report (2017). https://www.opnfv.org/
69. Specifications Technical Report Docket, ET., Technical report (2017). https://portal.3gpp.org/desktopmodules/specifications/specificationdetails.aspx?specificationid=3144
70. P. Neves, R. Calé, M. Costa, G. Gaspar, J. Alcaraz-Calero, Q. Wang, J. Nightingale, G. Bernini, G. Carrozzo, Á. Valdivieso et al., Future mode of operations for 5G–the SELFNET approach enabled by SDN/NFV. Comput. Stand. Interfaces **54**, 229–246 (2017)
71. C. Price, S. Rivera et al., OPNFV: an open platform to accelerate NFV. White Paper. A Linux Foundation Collaborative Project (2012)
72. OTF, Project boulder: intent northbound interface (NBI). Open Netw Found. **1**, 1–6 (2015)
73. P. Borril, M. Burgess, T. Craw, M. Dvorkin, A promise theory perspective on data networks. arXiv preprint: 1405.2627 (2014)
74. OTF, Neutron developer documentation. http://docs.openstack.org/developer/neutron/

D. K. Sah is currently Ph.D research scholar (full-time) at the Indian Institute of Technology (ISM), Dhanbad, from the Department of Computer Science & Engineering. His research areas are Wireless Sensor Networks, 5G networks, Network Security, Speech and Signal processing.
Email Id: dksah.iitd@gmail.com

D. Praveen Kumar is currently Ph.D research scholar (full-time) at the Indian Institute of Technology(ISM), Dhanbad, from the Department of Computer Science & Engineering. He is a Former Assistant professor and placement coordinator at Bapatla Engineering College, Bapatla, AP & Sree Venkateswara College of Engineering, Nellore, AP and also worked as Assistant Professor at Sri Venkatesa Perumal College of Engineering and Technology, Puttur, AP. He completed a Masters' in Technology and a Bachelor's in Technology at JNTUA, Ananthapuramu with Distinction. He had completed applied courses such as the PGDCA from a Ministry of HRD-recognized institute, Technical Teacher Training from NITTTR, Chennai, Basics of Computers from NIIT, Big Data Analytics With Hadoop and RHadoop from Skill Subsist Impels Ltd. He has published in one international journal and four international conference proceedings for IEEE, ACM, and Springer. He has given FDP and seminars on Big data analytics & Technical Training for Placements in various engineering colleges in AP. He is a member of SWIDC, CSTA, UACEE, IAENG, IJCSIT, CSI, and Swecha. He had one year of research experience at Yogi Vemana University. Research areas are Machine Learning and Wireless Sensor Network.
Email Id: praveeniitism@gmail.com

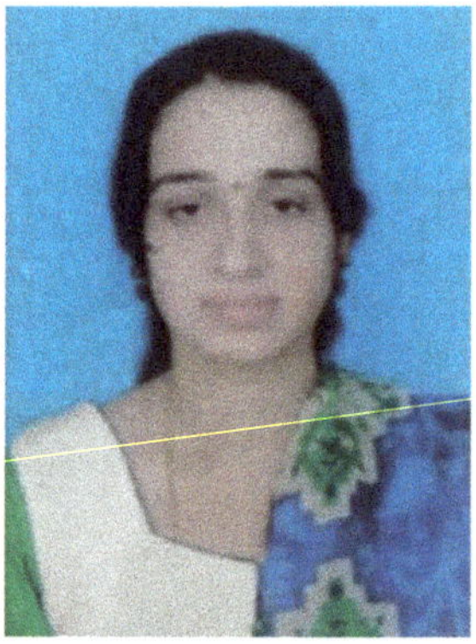

Chaya Shivalingagowda is currently Ph.D research scholar (part-time) at GITAM Vizag, from the Department of Electronics & Communication. She is Assistant Professor at Kalsekar Engineering College, New Panvel, Mumbai. She completed a Master's in Engineering at Mumbai University, Mumbai with Distinction. She has published in four international journals and two international conference proceedings for IEEE and Springer. She has given FDP and seminars on Open Source such as scilab, Ns2 Technical Training for Placements in various engineering colleges in Mumbai. She is a member of the UACEE, IETE. Research areas are Optical Communication and Wireless Sensor Network.
Email Id: chaya.ravindra@gmail.com

Dr. P. V. Y. Jayasree is currently working as Associate Professor at the Department of Electronics & Communication Engineering, GITAM university, Vizag Andhrapradesh, India-530045. She has published several articles in journals and conferences. She completed her Ph.D in Electronics and Communications Engineering with specialization in Electromagnetic Interference and Compatibility from JNTU Kakinada in July 2010. She completed her M.E in Electronics and Communications Engineering at Andhra University, Visakhapatnam in July 1999, and her B.E. in Electronics and Communications Engineering at the College of Engineering, GITAM, Andhra University in July 1989.
Email Id: pvyjayasree@gitam.edu

A Survey on the Security and the Evolution of Osmotic and Catalytic Computing for 5G Networks

Gaurav Choudhary and Vishal Sharma

1 Introduction

The next generation of wireless networks exponentially adopts various emerging technologies to facilitate the service with a high-speed data rate [1–5]. The 5G technology is adopted in various applications like Industrial-Internet of Things (IIoT), smart city, and smart grid, to fulfill the user requirements and fasten the speed with low latency and high reliability for remotely accessing the regular services [6–8]. The 5G setup aims at exploiting different types of nodes which ensure on-demand as well as reliable connectivity to other devices [9–11]. With a tremendous increase in the number of devices, the pressure of maintaining the quality as well as the quantity will increase which also puts considerable impact on the security policies of a network.

Different technology enablers such as blockchain, distributed mobility management, edge computing, osmotic computing, catalytic computing, or fog networks can help to sustain this pressure imposed by the security requirements. However, there is gap in the literature as there are not sufficiently evident solutions which can cover both the performance as well as the security aspects of 5G networks. Use of diversified sensor nodes, drones, and autonomous vehicles can further impose extensive challenge on using 5G services without being compromised [6, 7, 10, 11]. Thus, with the rapid development of technologies in the 5G era, security becomes a major concern for successful implementations. Moreover, threats and vulnerabilities have been evolving continuously with the networks because of an increase in the number of users where attackers try to exploit the potential weaknesses. To easily follow these issues and aspects, an overview of 5G applications, technologies, features,

G. Choudhary · V. Sharma (✉)
Department of Information Security Engineering, Soonchunhyang University, Asan-si, South Korea

D. N. K. Jayakody et al. (eds.), *5G Enabled Secure Wireless Networks*,
https://doi.org/10.1007/978-3-030-03508-2_3

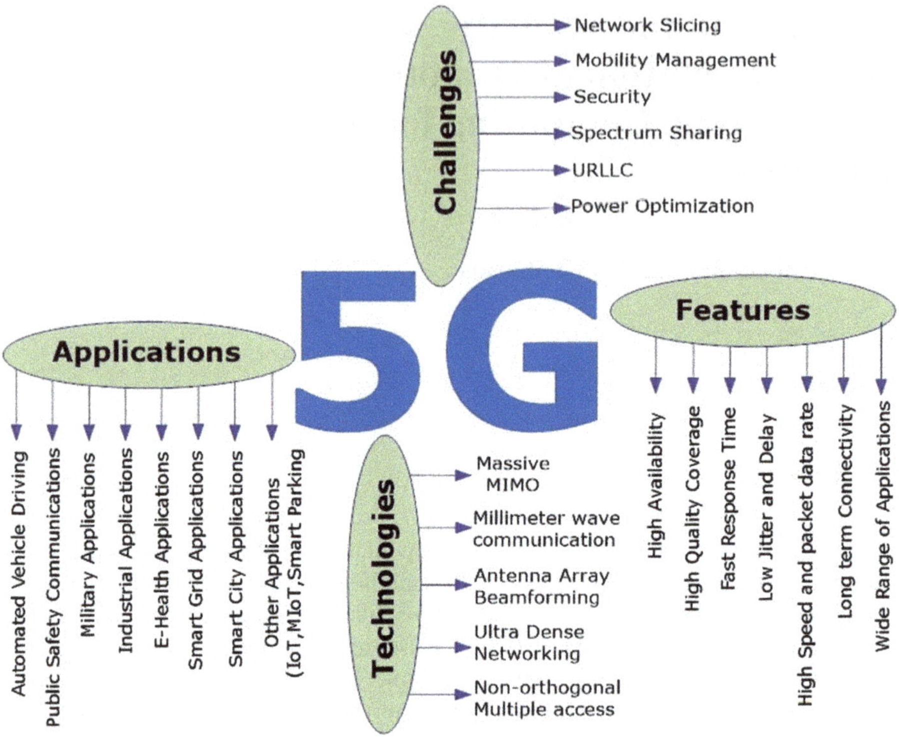

Fig. 1 An illustration of 5G applications, technologies, features, and challenges

and performance challenges and standardization is presented in Figs. 1 and 2. Current 5G networks are supported by different technology enablers, which include multiple-input multiple-output (MIMO), Massive-MIMO, non-orthogonal multiple access (NOMA), simultaneous wireless information and power transfer (SWIPT), orthogonal frequency division multiple access (ODFMA), radio access networks (RAN), network function virtualizations (NFV), software-defined networks (SDNs), device-to-device communications (D2D), network slicing, low-power wide area networks (LPWAN), etc. [12–21].

1.1 Applications of 5G Networks

The major applications of the 5G networks are provided below:

1. *Autonomous Vehicle*: The automatic controlled driving car and vehicles are key enablers of Vehicle-to-Vehicle (V2V), Vehicle-to-Infrastructure (V2I), and other Intelligent Transport Systems (ITS). The 5G network supports large bandwidth

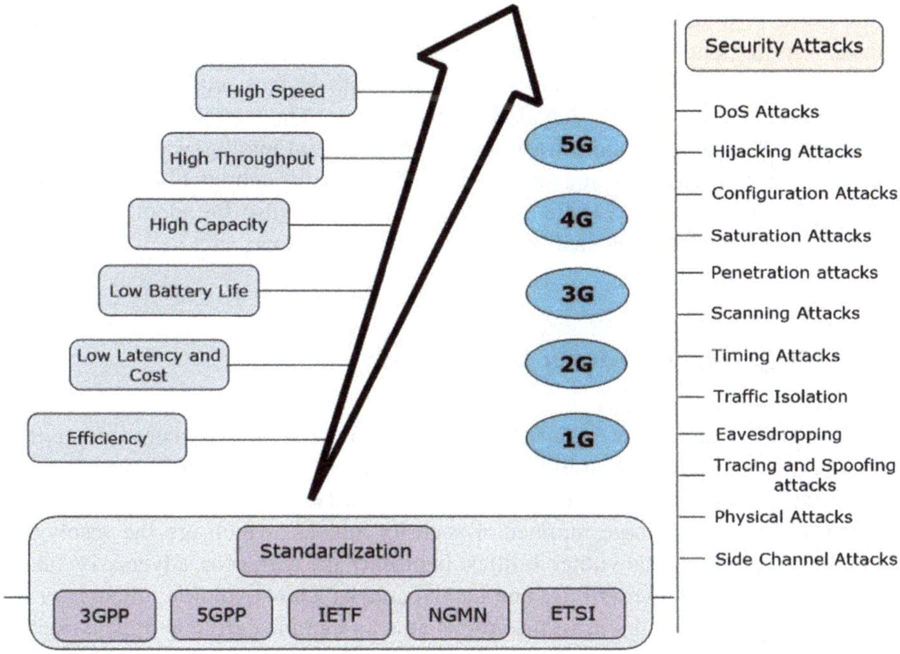

Fig. 2 An illustration of 5G evolution, standardization, and known security attacks

and low latency for these applications with high connection reliability. This network supports collision avoidance and intelligent navigation for the reliable transportation systems [6, 7].

2. *Public Safety Communications (PSCs):* The PSCs facilitated various communication services in case of emergencies when the primary communication infrastructure is not available. The PSCs incorporated rapid deployment and accessibility of communication setup. Therefore, 5G-enabled communications support a wide range of applications and long-term connectivity for the services in case of emergencies [12].
3. *Military Applications*: The military involved mission-critical control application which requires high data rate and long-term connectivity including the security parameters. The real-time surveillance and monitoring of suspected areas require a network with the large bandwidth and low latency. The 5G-enabled network application is the best fit for such mission-oriented applications.
4. *Industrial Applications*: Industrial automation composes massive IoT networks which require high connection density and low-power consumption that can be ensured through 5G setups [22–24].
5. *e-Health Applications*: The e-Health application requires remote diagnosis and long-term monitoring. The e-Health setup involves video streaming-embedded devices and advanced robotics which operate over the network that has

low-power, low-latency, and high-throughput requirements. The 5G facilitates these requirements and serve as the best solution for these applications.

6. *Smart City Applications:* Smart city adopts IoT devices, connected utilities, transportation, healthcare, education, smart grid, etc. These applications scenario requires automation, cloud infrastructure, and artificial intelligence which operate over a network composing large bandwidth, high throughput, high connection density, and low latency [10]. The 5G networks can also be adopted by new technologies and industry enablers like robotics and drones, etc.

1.2 Attacks and Threats in 5G Networks

Security holds a three-way relationship with the ease of deployment and user-friendliness. A more user-friendly setup often has poor security considerations and ease of deployment also reduces the level of security for any application. Such a trade-off results in a large number of security attacks, which are the result of intentional or unintentional vulnerabilities. In most of the scenarios, adversaries take advantage of known vulnerabilities and launch attacks while exploiting them over a period of time. Some of the most hazardous attacks include zero-day attacks, self-exploitable attacks, or side channel attacks [3, 15]. Along with these attacks, there a certain set of known attacks, as shown in Fig. 2, which impose a huge impact on the performance of any network. Considering the reachability and number of devices active in 5G networks, the scale and the impact of these attacks become severe and allow zero time to respond.

Efficient mutual authentication and strong key agreements should be followed while deploying applications in 5G networks; however, these should come without affecting the performance of the network [25–27]. Although 5G networks aim at increasing the speed as well as the quality of the transmissions, this reduces the time window for applying security mechanisms to prevent any active attacks on the networks. Thus, new mechanisms are needed for securing these fast communications or applicability approach of existing solutions needs to be changed to cope with the low-latency requirements of these networks [28–30]. Along with these attacks, it is desired the network should be free from insider threats as these may expose the internal functionalities which may lead to different types of attacks that are next to impossible to detect at the real time. Further, it is recommended to develop low-cost lightweight intrusion detection systems (IDS), which can fixate on these requirements and can help identify attacks prior to their launching.

2 Preliminaries: Osmotic Computing

Osmotic computing is a network integration paradigm inspired by the chemical process of osmosis. In chemistry, the osmosis is a spontaneous mechanism of transferring or movement of the solvent through a semipermeable membrane into solute concentrated solution. The solution is a mixture in which one substance is dissolved in another, the solute is the substance that is getting dissolved, and the solvent is the substance that is causing the dissolving. The standard mechanism of osmosis was molded with respective of available information by Villari et al. [31] to form this new paradigm of osmotic computing. This type of computing supports load balancing by enabling the movement of micro-services between a data center and the edge devices. It also reduces the latency of the overall application; however, in the initial drafts, the authors do not fully identify the mechanisms that can be used to support and facilitate such service migrations as stated in Sharma et al. [32].

As per the standard mechanism of osmosis, the model contains a solution of the solute and the solvent and a semipermeable membrane for filtering the solute. The participants of any model are interchanged with the respective terminologies to allow applicability of osmotic computing, like information and services, refer loads, process time, and energy of the system. The server can play the role of solvent for the interchangeable services through a semipermeable membrane and is configured as a controller for the movement of services between the senders and the receivers. The basic principle of osmosis is used for load balancing of incoming requests. In case a server receives a lot of requests and is unable to handle in those instances, the service provider can act as a semipermeable membrane and can shift the load to another server. To further enhance the understanding, the standard procedures without osmosis and load balancing with the osmosis principle are conceptualized in Fig. 3.

The primary objective of osmotic computing is to balance the load and resource utilization among servers without affecting the connectivity of services and performance. The real-time distribution of services as per the requirements helps to enable the new service migration concepts with the osmotic computing. Maksimović [33] defined the osmotic computing as "It enables the dynamic arrangement and migration of services and micro-services across cloud data centers and Edge resources according to different infrastructure demands and software." The author also conceptualizes the role of osmotic computing in the IoT and discusses the issues associated with the efficient execution of IoT services and applications across different computing infrastructures.

In the era of osmotic computing development, there have been limited but qualitative works that have highlighted the significance of using this new computing paradigm. Sharma et al. [32] focused on the efficient distribution and allocation of services via a concept of osmosis. The authors present a fitness-based osmosis algorithm which is used to provide support for osmotic computing. The algorithm utilizes a fitness function to distribute and allocate the services into micro- and macro-components. As per the authors, security concerns of the osmotic computing are still an open issue and future aspects should be considered with these issues. Furthermore, Sharma et al. [34] presented an efficient implementation of the osmotic computing, which is used for the pervasive trust management framework to perform computational off-loading. The authors used three solutions which include the models fitness-based movement, probabilistic movement, and threshold-based movement.

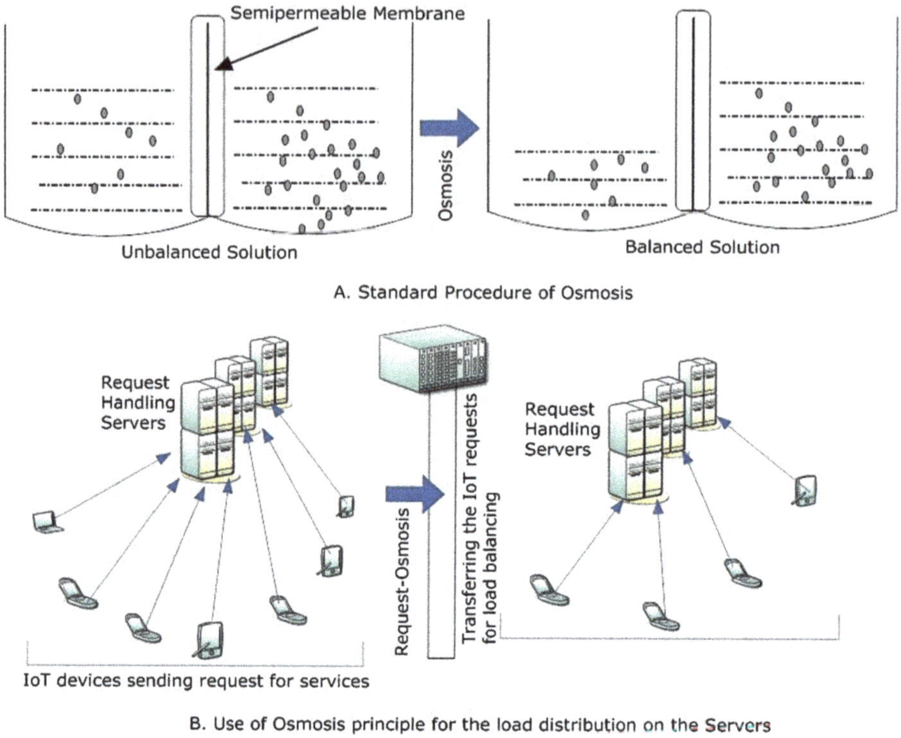

Fig. 3 The standard osmosis procedure and the use of osmosis principle for load balancing on servers

Nardelli et al. [35] presented the osmotic flow model, which supports multiple types and the mix of data transformation tasks on shared EDC+CDC infrastructures. The proposed model helps to assign the edge and cloud nodes among multiple and concurrent data transformation functions. As per the authors, the concept of machine learning with the distribution of data analysis tasks in various cloud environments imposes various solutions in a holistic manner. Morshed et al. [36] emphasized the deep learning concept in osmotic computing. The authors discussed the implementation issues and challenges of deep learning formations in the osmotic setup.

Rausch et al. [37] emphasized the applicability of osmotic computing to the message-oriented middleware for enhancing reliability, ultra-low-latency, and privacy-aware message routing. The proposed solution enables the facilities for brokers to enable or diffuse the edge resources on the demand and support device-to-device communication in the IoT environment. Despite the facilities associated with osmotic computing, lots of challenges are also linked with the emerging technologies. Buzachis et al. [38] addressed the issues of connectivity in the environment which imposes network degradation and failures concerning osmotic nodes.

3 Preliminaries: Catalytic Computing

Catalytic computing is obtained from the principle of catalysis. In chemistry, catalysis involves the addition of a substance or material to move forward a reaction, where the added substance or material gets used by components of the chemical reaction. In the catalysis procedure, the substance which plays a significant role in the acceleration of chemical reactions is known as catalysts. The principle of catalysis is used with a network for providing efficient resource sharing without affecting the operations of other components and without concerning its own performance. As given in [39], the catalytic computing can be considered as a new paradigm to provide efficient resource sharing in wireless networks comprising users with high mobility. The authors used a homogeneous discrete Markov model for user mobility patterns to decide on the applicability of catalytic computing, selection of catalyst, and procedures to fixate the activation energy. Furthermore, the authors present future aspect of newly coined paradigm for solving the problems like multi-network optimization problems, spectrum-sharing decision systems, scheduling problems, priority-based swarm communication, etc.

The standard process of catalysis and catalytic computing paradigm are illustrated in the Fig. 4. The catalyst in catalytic computing has the most important role to play, and it operates between the entities allowing efficient management and utilization of network resources. The network catalysts are the supporting entities which facilitate the network activity and allow zero-downtime services to the users by acting as a decision system. The catalytic computing obeys the properties of the catalysis process and supports continuous decisions through centralized or distributed operations.

4 Existing Surveys and Their Applicability

Over the last few years, various surveys have been published on the security issues and the challenges of 5G networks. Chopra et al. [40] discussed the security issues of the physical layer in Massive-MIMO, jamming, vehicular ad hoc network (VANET), and D2D along with challenges and solutions for 5G ultra-dense wireless networks. The authors find the attacking regions in the ultra-dense networks. The authors also presented the types of jammers and their countermeasures. Ferrag et al. [41] discussed the authentication and privacy-preserving schemes in 5G. The authors presented the classification of the threat models and respective countermeasures against the attacks. Fang et al. [42] discussed the recent development of security solutions in 5G. The authors included a discussion on the security features with the recent technologies focusing heterogeneous networks (HetNet), D2D, Massive-MIMO, SDN, and IoT. The authors also paid attention to the 5G

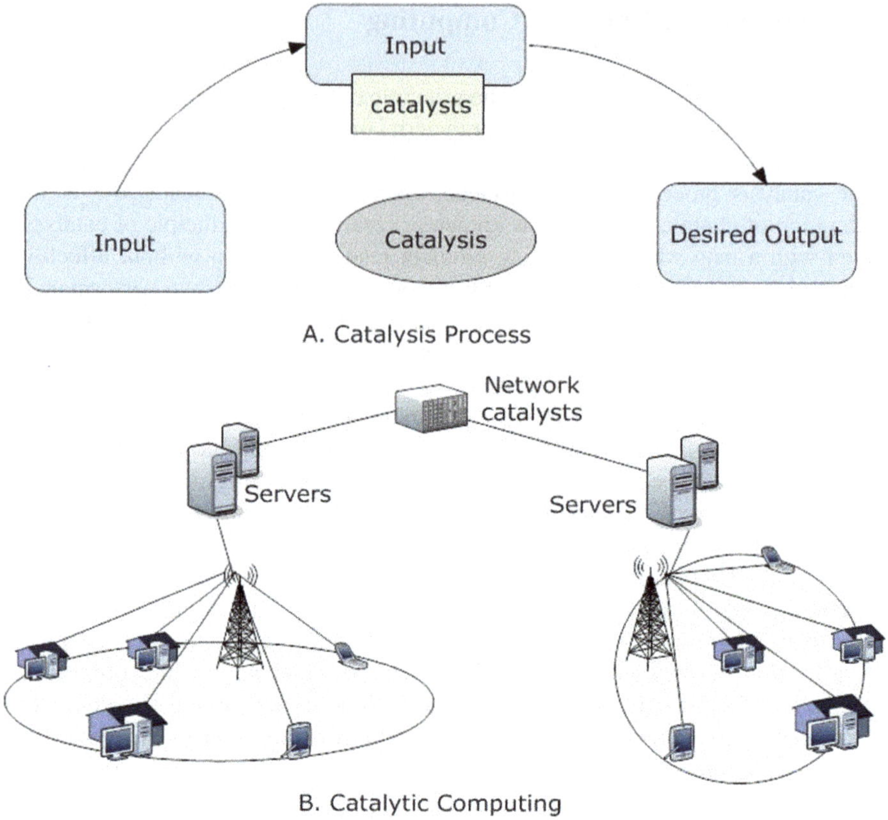

Fig. 4 An exemplary illustration of the standard catalysis process and inspired catalytic computing for networks

wireless security services like identity management and flexible authentications. Furthermore, the security services are also analyzed on the basis of the proposed architecture.

Lin et al. [43] presented the taxonomy of the network security-related data collection technologies. The authors analyzed the data collection nodes, mechanism, and tools. The authors also proposed the objective and requirements for the security-related data collections in a heterogeneous network. Furthermore, the research issues and future research challenges are considered in their survey. Rupprecht et al. [44] emphasized the main root cause of attacks and their defense in the mobile networks. The authors presented the challenges and research directions for the 5G security. The authors highlighted the limitations and drawbacks of existing works and security requirements for new 5G technologies.

The attacks and vulnerabilities of attacks, like DoS, are considerable issues for the future mobile generations. Ahmad et al. [45] presented the review of the challenges and privacy and security issues in the clouds, software-defined networking, and network function virtualization along with their perspective security solutions. The major contribution of the authors includes the analysis of the attack vulnerabilities and mitigation for making a secure 5G network.

The physical layer security approaches are robust and help to protect against eavesdroppers and man in the middle attacks. In addition, these provide flexibility for the secret key generation in 5G networks. Wu et al. [21] presented the survey on the physical layer security technologies Massive-MIMO, mmWave communications, heterogeneous networks, NOMA, full duplex technology, and challenges associated with these technologies. Furthermore, Gau et al. [46] gave a study of the physical layer security in the 5G-based social networks. The authors discussed the challenges of the physical layer security in the 5G-based large networks. The authors presented the solutions like security enhancement in the system layer, link layer, and cross-layer optimization.

Some more kinds of studies have been presented in Ji et al. [47] who analyze the security requirements of 5G business applications, network architecture, the air interface, and user privacy and in Panwar et al. [48] who discussed the limitations of 4G networks and the features of 5G networks. The authors discussed the challenges, technologies, and implementation issues of these networks. Le et al. [49] presented the technology enablers for 5G communication and security challenges of the emerging technologies. From these studies, it can be marked that the security is a significant constraint for the 5G architecture, and to facilitate the reliable services on the mobile communication, it is necessary to consider privacy and trust of the user. The details of comparative evaluations and roadmap of existing surveys are presented in Table 1 and Fig. 5.

5 Taxonomy of Security Concerns for 5G Networks

This section discusses the five major paradigms identified for the security of 5G networks as shown in Fig. 6. The section provides an overview of existing approaches along with their comparative evaluations.

5.1 Secure Resource Allocation in 5G

In wireless communications, resource allocation plays a significant role. The resource allocation is defined as the management of resources and increases the capacity for a better communication. The various resources like power, load, spectrum, etc. are shared according to the user requirement for an optimized connectivity. The resources in the cellular networks can be assigned locally or

Table 1 A comparative overview of the existing surveys on the security of 5G networks

Survey	Year	Application area	Key contribution
Chopra et al. [40]	2017	Ultra-dense network	Discuss security issues, challenges, and respective solutions for 5G ultra-dense wireless networks
Ferrag et al. [41]	2017	Cellular networks	Presents the classification and comparisons of the authentication and privacy-preserving schemes, threat models, and respective countermeasures against the attacks
Fang et al. [42]	2018	5G mobile wireless networks	The authors discuss the recent development of security solution in the era of 5G. The author includes discussion on the security features with the recent technologies like HetNet, D2D, Massive-MIMO, SDN, and IoT
Lin et al. [43]	2018	Network security	The authors present the taxonomy of the network security-related data collection technologies. The authors analyze the data collection nodes, mechanism, and tools
Rupprecht et al. [44]	2018	Mobile network generations	The authors discuss the root cause for attacks and their defense. The authors discuss the challenges and research directions on the 5G security
Ahmad et al. [45]	2018	5G networks	The authors review the challenges and issues of privacy and security in the clouds, software-defined networking, and network functions virtualization and presents respective solution for these issues
Wu et al. [21]	2018	5G wireless networks	The author presents the survey on the physical layer security techniques, challenges associated with these technologies, and future trends
Ji et al. [47]	2018	5G networks	The authors analyze the security requirements of 5G business applications, network architecture, the air interface, and user privacy
Panwar et al. [48]	2016	5G mobile communication	The authors discuss the limitations of 4G and features of 5G. The authors discuss challenges, technologies, proposed architecture, and implementation issues
Le et al. [49]	2016	5G networks	The authors discuss the technologies which enable 5G mobile communications. The authors emphasized on the challenging issues and future directions in 5G mobile access networks

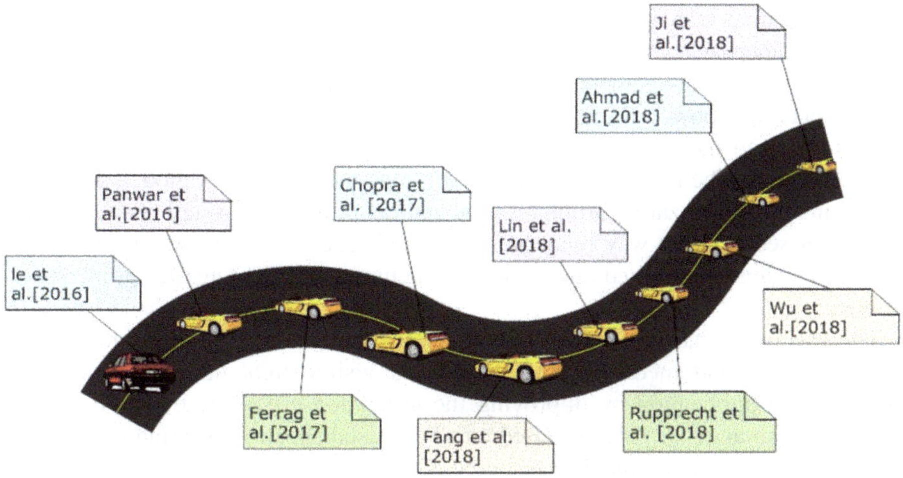

Fig. 5 The roadmap of existing surveys on the security descriptions of 5G networks

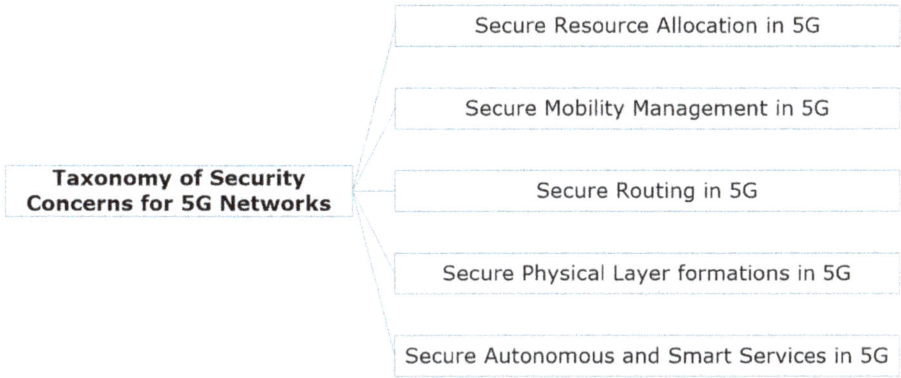

Fig. 6 An illustration of classification of security concerns in 5G networks

globally. The traditional solutions emphasized the research allocations strategies by incorporating mathematical game theory problems but do not put a compatibility with security facilities. The security among resource sharing is essential from the perspective of optimization and better use of resources. The eavesdropper or an attacker can use the resources in a falsified way and increase the network degradation rate. Therefore, the secure resource allocation in the 5G networks is a crucial challenge and should be focused for improved connectivity.

Various studies have been presented in the era of secure resource allocation in 5G networks. Yu et al. [50] presented the optimal resource sharing scheme on the basis of the matrix game theory. The user adopts the concept of enhanced cloud radio access network (EC-RAN) for supporting data-heavy applications in the vehicular networks in a 5G communications environment. The authors also discussed the cloudlet resource management approach. Luong et al. [51] focused the economic and pricing approaches for resource management in the 5G networks. The authors use the resource management issues like user association, spectrum allocation, interference, and power management to define appropriate models. Zhang et al. [52] emphasized the secure resource allocation for orthogonal frequency division multiple access. The mixed integer programming and non-convex models are used for solving the optimization problems. The authors proposed a system model on the concept of secrecy two-way relay WSNs without cooperative Jamming. Abedi et al. [53] presented the limited rate feedback scheme by formulating the optimization problem.

The spectrum sharing enables various unlicensed bands, including the industrial, scientific, and medical (ISM) band, and visible light communication for the dedicated application by improving the utilization efficiency of the available spectrum. Akhtar et al. [54] proposed an SDN-based spectrum-sharing technique. Liu et al. [55] discussed the 3-D resource allocation techniques. For the sum rate maximization, the authors combined the antenna selection and the user scheduling algorithms. Yang et al. [56] presented the random-based radio resource allocation approach for D2D communications. In addition, network slicing can be a considerable solution for enhancing the capacity and flexibility of the networks in secure resource allocation and management schemes. Tremendous adoption of the 5G network with different application facilitates massive traffic and data on the networks [57]. The details of comparison between the state-of-the-art solutions for secure resource allocation in 5G Networks in presented in Table 2.

5.2 Secure Mobility Management in 5G

The 5G network imposes a high data rate transmission in the communication and delivery of services. The high data rates can be achieved through better traffic control and mobility management. Various technologies incorporated handover as a key concept for mobility management. For the better communication in the 5G scenario, various researches have been conducted for mobility management with the security consideration of the communications. In the era of mobility management in 5G, Sharma et al. [39] presented a resource-based mobility management scheme. The proposed solution is based on the catalytic computing. The authors used a homogeneous discrete Markov model for analyzing mobility patterns of the user and congestion control algorithm. The handover mechanism is presented on the basis of activation energy.

Furthermore, for the handover schemes for the 5G networks, Sharma et al. [71] presented a secure key exchange and authentication protocol for fast handover in the 5G Xhaul networks. The proposed protocol provides security over the links for the moving terminals in the networks. The authors discussed the security of the backhaul, fronthaul, and Xhaul. You and Lee [72] gave a ticket-based handover approach for the 5G networks. The proposed mechanisms rely on the concept of

Table 2 State-of-the-art solutions for secure resource allocation in 5G networks

Approach	Author (s)	Ideology	R1	R2	R3	R4	R5	R6	R7
Optimal resource sharing	Yu et al. [50]	Matrix game theoretical approach	Yes	Yes	No	Yes	No	No	No
Economic and pricing approaches for resource management	Luong et al. [51]	Resource management issues like user association, spectrum allocation, interference, and power management	Yes	Yes	Yes	Yes	No	No	No
Secure resource allocation	Zhang et al. [52]	Mixed integer programming problem	–	–	Yes	Yes	No	No	No
Limited rate feedback scheme	Abedi et al. [53]	Formulate as an optimization problem	–	–	Yes	Yes	No	No	No
HAS: harmonized SDN-enabled approach	Akhtar et al. [54]	Centralized management based on distributed inputs	–	Yes	Yes	Yes	No	No	No
5G network slice broker	Samdanis et al. [58]	Logically centralized monitoring and control entity	Yes	Yes	No	Yes	No	No	No
Radio resource allocation method for D2D communication	Yang et al. [56]	Random-based approach	No	No	No	Yes	No	No	No
3-D resource allocation techniques	Liu et al. [55]	Combine antenna selection and user scheduling	No	No	No	Yes	No	No	No
Architecture for network slicing-based 5G systems	Zhang et al. [57]	Based on SDN and NFV technologies	Yes	Yes	–	Yes	No	No	No
Resource allocation for secure communication	Wang et al. [59]	Three-dimensional stochastic model	Yes	–	Yes	Yes	No	No	No
Joint resource allocation framework	Li et al. [60]	Heuristic VNE algorithms	No	Yes	–	Yes	No	No	No
Green resource allocation scheme	AlQerm and Shihada [61]	Centralized and decentralized sophisticated online learning scheme	No	No	No	Yes	No	No	No
Secure communications in NOMA system	Zhang et al. [62]	Formulate in the non-convex optimization problem	No	No	Yes	Yes	No	No	No
Hierarchical radio resource management scheme	Belikaidis et al. [63]	Spectrum access system	–	No	No	Yes	No	No	No
Network resource allocation system	Martin et al. [64]	Machine learning methods	Yes	Yes	No	Yes	No	No	No

(continued)

Table 2 (continued)

Approach	Author (s)	Ideology	R1	R2	R3	R4	R5	R6	R7
Social trust scheme	He et al. [65]	Deep learning approach	–	–	Yes	Yes	No	No	No
Framework of resource allocation and performance optimization	Tan [66]	Based on control theory	Yes	Yes	Yes	Yes	No	No	No
Network slicing management and prioritization	Jiang et al. [67]	Heuristic-based admission control mechanism	No	No	No	Yes	No	No	No
Energy-efficient resource allocation	AlQerm and Shihada [68]	Machine learning scheme	No	No	No	Yes	No	No	No
SDN-based resource management	Duan [69]	Programmable management platform	–	Yes	Yes	Yes	–	No	No
Resource allocation scheme	Zhu [70]	Geometric approach that analyzes the feasible region governed by the constraints	–	–	Yes	Yes	–	No	No
Resource-based mobility management	Sharma et al. [39]	Catalytic computing	No	Yes	No	Yes	No	No	Yes
Service management scheme	Sharma et al. [32]	Osmotic computing	No	Yes	No	Yes	No	Yes	No

R1 reliability, *R2* low latency, *R3* security, *R4* resource allocation/scheduling, *R5* key management, *R6* osmotic principles, *R7* catalytic principles reducing messages involved in the authentication procedure with the authentication server. The user anonymity in the handover is considered by Fan et al. [73] in their ReHand scheme. Their scheme reduces the communication cost compared with the existing solutions.

Munir et al. [75] presented a secure fault tolerance mechanism for 5G networks based on the concept of the distributed hash table of access nodes and ticket-reuse approach. The authors presented an analytical model for analyzing the processing time of a location query with the minimum authentication latency. To support distributed mobility management (DMM), Sharma et al. [76] developed a blockchain-based DMM handover scheme. Their solution focused on the resolution of hierarchical security issues without affecting the network layout. Some other solutions can be studied from refs. [77–80]. In short, 5G should be adopted with a reliable and scalable architecture for the mobility management to support a high data rate with the minimum delays and latency. The details of comparison between the state-of-the-art solutions for secure mobility management in 5G networks are presented in Table 3.

5.3 Secure Routing in 5G

The 5G networks enhance the connectivity of various applications with the high speed, low latency, and high availability. The standard architecture considers various network management challenges. One of the major considerations in these challenges is the secure routing. With the increase in traffic, it is very difficult to manage services and data in a real-time along with their secure deployments. The standard network management mechanisms pay attention to the routing and monitoring of data flow. The data transmission and routing should be protected against various kinds of attacks like the man in the middle, downgrading attacks, location spoofing, and resource depletion attacks.

For the efficient and reliable data communication in the 5G era, secure routing is an emerging research area. Over the years, an extensive research has been conducted on the secure routing in the 5G networks. Guan et al. [81] presented a GRBC-based network security functions placement scheme with applicability to the software-defined security. The authors focused on the loop-free routing schemes and independent routing decisions. Sharma et al. [39] discussed the routing-based mobility management, SDN-based mobility management, and cluster-based mobility management mechanisms. The routing-based mobility management is used in the older concept with the benefit of the optimal path for traffic flow. The proposed approach, based on the catalytic computing, supports the user movement control and routing-based handovers.

Jung et al. [82] presented a joint operation protocol for the group key management and routing. Routing control helps to maintain the change in state of the links. Wang and Yan [83] discussed various secure routing schemes like the Secure Message Delivery (SMD) protocol and Secure Optimized Link State Routing

Table 3 State-of-the-art solutions for secure mobility management in 5G networks

Approach	Concept	Parameters used	R1	R2	R3	R4
Key exchange and authentication protocol [71]	Securing Xhaul links for a moving terminal in the network	Failure factor, packet loss, signaling overheads	Yes	Yes	Yes	No
SPFP: ticket-based secure handover [72]	Decrease the number of message related to authentication	Handover latency, packet loss/buffering, and handover failure	Yes	Yes	Yes	No
ReHand: secure region-based handover scheme [73]	Incorporate the techniques of group key, one-time identity, and one-way hash	Computational cost, communication cost	Yes	Yes	Yes	No
User-centric ultra-dense networking [74]	Dynamic AP grouping	Solutions for mobility management and mobility scenario	Yes	Yes	Yes	No
Secure and fault-tolerant mechanism for DMM [75]	Used distributed hash table of access nodes and ticket-reuse approach	Average authentication latency, average processing time	Yes	Yes	Yes	No
Blockchain-based DMM [76]	Resolving hierarchical security issues without affecting the network layout	Energy consumption	Yes	Yes	Yes	No
Mobility and traffic management mechanism for delay tolerant cloud data [77]	Selecting appropriate cell for data transfer in downlink, and by configuring the UE cell priorities for sending the data in uplink	Throughput distribution, delay	No	No	No	No
Policy-based communications [78]	System controlled by policy to detect misbehavior	Session setup delays	No	No	No	No
Policy-based per-flow mobility management system [79]	The decisions are made on the basis of policies	Handover decision making	Yes	No	No	No
Secure producer mobility management [80]	Use concept of hash-chain that protects against prefix hijacking attacks occurring during mobility updates	Storage cost, edge router goodput, verification delay	Yes	Yes	–	No
Resource-based mobility management [39]	Catalytic computing	Equilibrium evaluation, activation energy	Yes	No	–	Yes

R1 handovers, *R2* key management, *R3* mutual authentication, *R4* osmotics/catalytic principles protocol (SOLSR) that have a functionality to protect the messages relayed between the source and the destination. The authors also analyzed the secure routing, access control, and physical layer security requirements for D2D communications.

Choyi et al. [84] focused on the hybrid routing through network slicing. The authors also discussed the static and the dynamic mechanisms of routing. The static and the functional overviews of the service negotiation framework and flexible routing with network slicing are also discussed. In their framework, routing is used by the static path defined by packets. Naqvi et al. [85] presented the study for the acceptance possibilities of the IPv6 within 5G wireless networks. The authors presented a study of inter-domain routing and the comparison study with the IPv4.

Schmittner et al. [86] focused the PSCs and presented a novel secure multi-hop D2D scheme. The proposed scheme is used for forwarding decision that is based on reliability metric for the routing. Zhao et al. [87] presented a cluster-based technique focusing security-based transmission protocol for ultra-dense networks (UDNs). The authors analyzed the security issues and resource management schemes for better network formations. Furthermore, a secure routing protocol is presented for the multi-hop networks. The protocol is based on the concept of Weil paring [88]. The secure routing enables a better communication and service management via traffic flow and optimized use of resources. Therefore, the secure routing mechanism should be included in the 5G network for a reliable and secure communications. The details of comparison between the state-of-the-art solutions for secure routing in 5G networks are presented in Table 4.

5.4 Secure Physical Layer Formations in 5G

The physical layer security gains a hike in the 5G network communications. The conventional cryptographic solutions for security impose the difficulties in the distribution and management of secret keys. But the physical layer has provided reliable and flexible security levels through various protocols. The concept of the physical layer is to "utilize the intrinsic randomness of the transmission channel to guarantee the security in physical layer" [21]. The existing approaches to security require strong constraints and high additional costs for the users of public networks. Therefore, the new concept on physical layer security focuses on the secrecy capacity of the propagation channel, which is adopted by the 5G networks.

Gomez et al. [89] analyzed the physical layer security for uplink NOMA in 5G large area networks. The key mechanism of coverage probability and the effective secrecy throughput is used in their solution. The signal-to-interference plus noise ratio (SINR) is measured with the legitimate user and attackers. Furthermore, Forouzesh et al. [90] analyzed the physical layer security for Power Domain (PD)-NOMA in HetNet. The novel concept of resource allocation scheme is presented to maximize the sum secrecy rate in PD-NOMA-based HetNet. The interference in the signals is considered as an eavesdropper in their approach. An alternative search method algorithm is adopted for the optimization problem. Furthermore,

Table 4 State-of-the-art solutions for secure routing in 5G networks

Approach	Mechanism	Parameters used	R1	R2	R3	R4
Catalysis-based mobility management [39]	Homogeneous discrete Markov model for user mobility patterns	Equilibrium evaluation, activation energy	No	No	No	Yes
Group routing betweenness centrality [81]	Finding a group of given size with maximum GRBC	Complexity analysis, network topology, and attack model	No	No	No	No
Joint operation protocol [82]	Use group key agreement method	Computational load, latency time	Yes	No	Yes	No
Secure routing [83]	Secure Message Delivery (SMD) protocol	D2D security architecture and security requirements	Yes	Yes	Yes	No.
Network slices and routing [84]	Hybrid routing	Packet routing	No	No	No	No
IPv6 technology [85]	OPNET MIPV6 model	Throughput, network delay, and packet delay	No	No	No	No
SEMUD [86]	Per-destination routing state used	Throughput, goodput	No	No	No	No
Cluster-based UDNs [87]	Security-based transmission protocol	Network eavesdropping defense and jamming attack defense	Yes	No	No	No
Secure routing protocol [88]	Weil paring based	End-to-end delay, throughput	Yes	_	Yes	No

R1 encryption, *R2* mutual authentication, *R3* key management, *R4* osmotics/catalytic principles

Liu et al. [94] discussed the physical layer security for 5G-NOMA on the basis of stochastic geometry approaches and the location of the user. Asymptotic secrecy outage probability is used for the multiple antennas, and the simulation result is verified with the Monte Carlo mechanisms.

Massive-MIMO has the capability to handle 100+ antenna elements and a large number of independent transceiver chains. The increasing adaptability of Massive-MIMO opens the feasibility of adaption to physical layer security against the eavesdroppers. Kapetanovic et al. [91] presented the opportunities and the challenges of physical layer security for Massive-MIMO. The authors presented various possible attack methods and detection of the active attacks on the Massive-MIMO. Wang et al. [92] discussed the physical layer security in cellular HetNet. The authors proposed a threshold-based secrecy mobile association policy and calculated secrecy and connection probabilities of the random users. The closed-form expressions are used with the secrecy and connection probabilities.

Zhang et al. [95] presented the cooperative anti-eavesdropping techniques on the basis of graph theory. In their approach, on the basis of network topology, a secrecy weighted graph is calculated. The proposed techniques facilitate the low complex security on the large-scale networks. Chen et al. [96] analyzed the physical layer security for cooperative NOMA systems. Pan et al. [97] presented hierarchal physical layer security architecture for the 5G networks based on the cross-layer lightweight authentication scheme and a physical layer security-assisted encryption scheme. These schemes are based on the key streams and the channel information. An overview of the state-of-the-art solutions for the physical layer security technologies in the 5G secure communications is presented in Table 5.

5.5 Secure Autonomous and Smart Services in 5G

Various architectures for autonomous services have been discussed that support the flexibility, performance, cost, security, safety, manageability, etc. [100]. Cheng et al. [101] discussed the architecture of 5G-based IIoT in the three application modes, namely, enhance mobile broadband (eMBB), massive machine-type communication (mMTC), and ultra-reliable and low latency communication (URLLC). The authors presented the applicability of the IIoT and cyber-physical manufacturing systems (CPMS) in the 5G wireless communication network and also discussed the key challenges in the adaptability of these networks.

Mavromatis et al. [102] presented architecture for connected and autonomous vehicles. This architecture relies on the mechanism of multilayer application data streaming. The authors presented the next-generation ITS for intelligent traffic planning, smart emergency vehicle routing, and multimodal commuting. Han et al. [104] discussed the mobile sensing and cloud computing and presented a combined concept of mobile cloud sensing.

The low latency applications in 5G networks have emerged stronger and are adopted widely for the business perspective. Lema et al. [105] discussed the

Table 5 State-of-the-art solutions for secure physical layer formations in 5G networks

Approach	Mechanism	R1	R2	R3	R4	R5
Physical layer security for NOMA [89]	Coverage probability and the effective secrecy throughput	5G, NOMA	Yes	No	No	No
Physical layer security for PD-NOMA [90]	Used alternative search method algorithm	5G, NOMA	Yes	No	No	No
Physical layer security for Massive-MIMO [91]	On the basis of pilot contamination scheme	MaMIMO, 5G	Yes	No	Yes	No
Physical layer security in HCN [92]	On the basis of access threshold-based secrecy mobile association policy	5G	Yes	No	No	No
Physical layer security for 5G-NOMA [93]	On the basis of stochastic geometry approaches, the location of user obtained	5G, NOMA	Yes	No	No	No
Physical layer security for NOMA [94]	Added eavesdropper exclusion area (protected area)	5G, NOMA	Yes	No	No	No
Cooperative anti-eavesdropping techniques [95]	On the basis of graph theory	P2P, D2D, 5G	Yes	No	No	No
Physical layer security for cooperative NOMA systems [96]	Secrecy outage probability (SOP) and strictly positive secrecy capacity (SPSC)	5G, NOMA	Yes	No	No	No
Hierarchical security architecture [97]	On the basis of cross-layer light weight authentication scheme	eMBB, mMTC, and URLLC,5G	No	No	No	No
Robust beam forming [98]	On the basis of robust information and artificial noise (AN) beam forming	Massive-MIMO, 5G, BDMA	No	No	No	No
Hybrid MIMO phased-array time-modulated directional modulation scheme [99]	Time-modulated DM scheme is added for the phased-MIMO to achieve PLS	MIMO, 5G	Yes	No	No	No

R1 technology used, *R2* secrecy, *R3* channel coding; *R4* jamming, *R5* osmotic/catalytic framework applications that rely on the low latency in the area of healthcare, automotive and transport systems, entertainment, and manufacturing from the market perspectives and respective models. Saghezchi et al. [106] emphasized the smart grid for smart cities and presented secure network architecture on the basis of detection of price integrity or load alteration attacks. Furthermore, Arfaoui et al. [107] presented the security architecture for 5G networks. From these studies, it is conclusive that the adaptable architecture for autonomous services in the 5G networks should provide flexible solutions and benefits for industries and leverage the new technology enablers for the overall development, especially security features, as a perspective of reducing the research and deployment cost. The details of comparison between the state-of-the-art solutions for secure services in 5G networks are presented in Table 6.

6 CATMOSIS: A Generalized Model for 5G Security

Both the osmotic and the catalytic computing can be united together to form an efficient security module which can be used by all types of applications. One such example is shown in Fig. 7. The figure shows four different catalytic managers for four different types of applications each of which are handled through the osmotic servers for cross-platform exchange of information. The intra-mode security of each application can be obtained through traditional mechanisms; however, with the amalgamation of osmotic and catalytic computing, it becomes convenient to secure the cross- or inter-platform services. This generalized model operates in two parts to form a CATMOSIS module for security as given below:

- *Catalytic Manager*: The module is responsible for handling all the intra-mode services and generate dependencies for secure communications. This module is similar to the one proposed by the original authors. It uses activation energy-based resources to decide on the policies and security requirements which are to be shared over the network platform.
- *Osmotic Manager*: This module is responsible for fetching the services to and from the catalytic manager while keeping intact the general flow of traffic. It helps to distinguish the security requests from the general traffic and allows categorization feature for providing enhanced security. However, the security is enhanced depending on the protocol or the methodology opted for authentication of devices as well as services. At the moment, flow authentication and categorization can be attained through this module, and other possibilities of extending this are left to future works.

Table 6 State-of-the-art solutions for secure services in 5G networks

Approach/architecture	Mechanism	R1	R2	R3	R4	R5
5G wireless communication technology [101]	On the basis of technical scheme and features of 5G wireless communication technology	5G, eMBB, mMTC, and URLLC	Yes	No	–	No
Architecture for connected and autonomous vehicles [102]	Multilayer application data streaming	5G, multiple radio access technologies	No	No	Mobility-as-a-Service (MaaS)	No
5G network architecture [103]	Intelligent use of network data	5G, Massive-MIMO, NFV, and SDN.	Yes	No	IaaS and XaaS	No
Architecture of mobile cloud sensing [104]	Mobile sensing and cloud computing	5G, Opinion Finder, Google Profile of Mood States (GPOMS)	No	No	All-IP network (AIPN) model	No
Use case with ultra-low latency in 5G [105]	Market perspectives of each industry and respective models	5G, advanced imaging, data analysis, and machine learning	Yes	No	B2B model	No
Secure network architecture for smart grids [106]	On the basis of detection of price integrity or load alteration attacks	5G, AMI, M2M	Yes	Yes	–	No
Security architecture for 5G networks [107]	Inherited concepts from the security architectures of 3G and 4G networks	5G, ITU-T X.805, 3GPP	Yes	No	Trust model	No
Fog computing [108]	Based on the TSI- NFV MANO architecture	5G, Peer-to-Peer (P2P)	Yes	Yes	–	No
Architecture for monitoring services [109]	Self-protection mechanism through monitoring	5G, SDN, NFV	Yes	Yes	Business support systems	No
Fog-based anomaly detection approach [110]	Unsupervised clustering and outlier detection algorithms	5G, LPWAN, edge computing, SDN, NFV	Yes	No	–	No

R1 technology used, *R2* security, *R3* IDS formations, *R4* business models, *R5* osmotic/catalytic framework

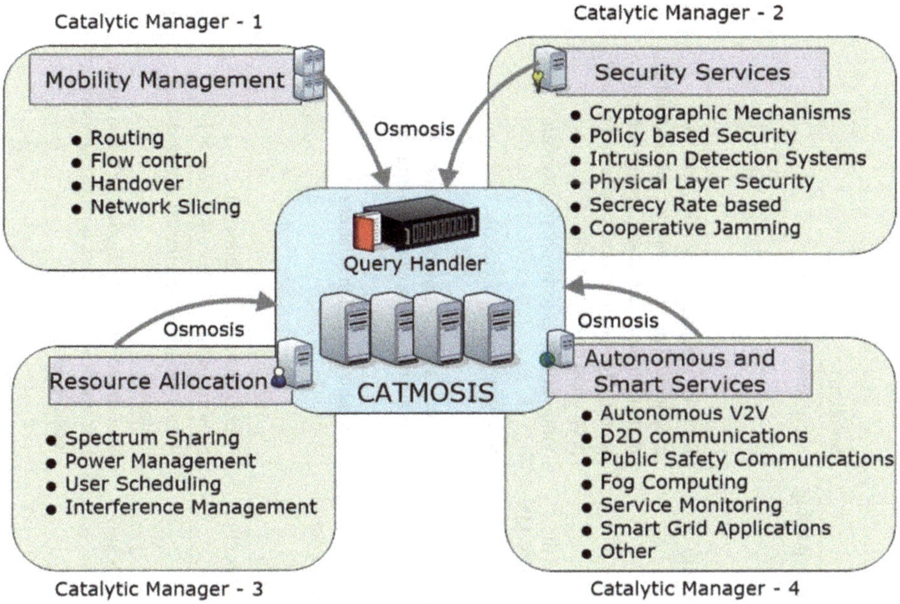

Fig. 7 An illustration of the CATMOSIS general model for security enhancement of 5G networks

7 Open Issues and Future Directions

5G era is focusing on the enhancement of applications which otherwise are difficult to attain at current network functionalities. Some of the other works to follow for enhancing the understandings on the 5G and its security are discussed in Table 7. Features like virtual reality networks, connectivity to billions of IoT devices, and information on the go are some of the major enablers in 5G setup [128]. However, with applications gaining new heights, security becomes primary as well as tedious objective to attain [129]. Thus, following these requirements, this article summarizes major security directions to be followed for further research as provided below:

1. *Secure and flexible architecture*: The standard architecture should be formed to support generalized applications from different domains. The flexibility in terms of adoption of various add-on features with the compatibility of the application perspectives is needed in the security architectures [130]. The security should be considered in different domains to handle attacks and security threats. Various secure architectures have been suggested to support the domain-specific application with the facilities of reliability and efficiency [131–133]. But from the research perspective, the adopted architecture should be enabled to follow generalized implementation mechanism within the application domains and sustain the security within the architecture.

Table 7 An overview of other studies to follow for the security of the 5G networks

Focus	Author (s)	Challenges and issues	Solutions
Analysis of the physical downlink and uplink control channels and signals	Lichtman et al. [111]	Jamming vulnerability, sniffing, and spoofing vulnerability	Mitigation techniques
5G security	Schneider and Horn [112]	Flexible security and potential security requirements	Security mechanisms for 5G- like user identity and device identity confidentiality, mutual authentication and key agreements, security between terminal and network, and protection against attacks
5G mobile communication networks	Luo et al. [113]	Traditional mechanisms does not find the potential vulnerabilities in the 5G-SDN-MN	Vulnerability assessment mechanism
5G vehicular networks	Ejiyeh and Talouki [114]	Attacks replay, message fabrication, and DoS attacks	Efficient security protocol
Network security in mobile 5G	Bouras et al. [115]	SDN security challenges: eavesdropping, identity spoofing, password-related attacks	General security guidelines, firewalls, administrative passwords, testing techniques, etc.
Security for 5G communications	Mantas et al. [116]	Potential threats and attacks for the following 5G system components: the UE, the access networks, the mobile operator's score network, and the external IP networks	Potential mitigation schemes
5G security	Svensson et al. [117]	Information leakage between network slices	Architecture for authentication, authorization, and accounting for 5G
5G wireless communication network	Gupta et al. [118]	Bandwidth spoofing attack	Attack modeling and intrusion detection system
5G security	Ahmad et al. [119]	Security threat, security issue in SDN and NFV and in communication channels	Security technology- DoS, DDoS detection, link verification, access controls, identity and location security, etc.

Network slicing in 5G	NGMN Alliance [120]	Issues: controlling inter-network slices communications, impersonation attacks, variation in the security protocols and resources exhaustion	Mutual authentication, integrity verification, baseline security levels for each slices, etc.
5G technologies	Felita and Suryanegara [121]	Discuss the technological challenges like security, limited frequency spectrum resources	Suggests: TRIESTE framework, BDMA technique
5G wireless communication networks	Yang et al. [122]	Focus on the security challenges in various technologies: HetNet, Massive-MIMO, and mmWave	Physical layer security
5G networks	Sun and Du [123]	Challenges for network security	Physical layer security approaches: artificial noise injection, anti-eavesdropping signal design, secure beam forming/precoding, etc.
Cloud computing in heterogeneous 5G	Gai et al. [124]	Crucial security concerns in mobile cloud computing	Intrusion detection techniques
5G cellular networks	Atat et al. [125]	Cellular transmissions protection and links protection from eavesdropping	Suggest security controls (authentication), monitoring of real-time data streams, implementing advanced anomaly detection techniques, etc.
5G network security	Mammela et al. [126]	Security visibility and configurability issues, threats, and vulnerabilities issues	Implementation of security monitoring spanning multiple micro-segments
5G wireless communication networks	Gandotra and Jha [127]	Secure power optimization issues	Standardization agencies and discuss on the ongoing projects

2. *Secure and continuous connectivity*: The massive increase in IoT devices in the 5G networks raises various connectivity issues. The connection should be continuous and stream less to facilitate real-time monitoring and service scheduling. The 5G network should facilitate security policies without affecting the services and connectivity of the network while maintaining the privacy of its users [134].
3. *Sustainability and reliability*: The application dependability is measured in terms of the reliability and the sustainability of the network. The resource depletion and unwanted service consumes a lot of resources over the network and impose an extra load on the network which results in sudden failures. The sustainability should be a considerable issue to resolve along with the efficiency and the reliability. The 5G-based applications require a wide range of the devices to connect with high connection density and low-power consumption, which require a focus on the reliability and the sustainability to enhance the network tolerance level [135, 136].
4. *Strong and efficient mutual authentications*: The security mechanisms in 5G should incorporate strong mutual authentication to verify the originality of both the receiver and the sender. The authentication mechanisms can be performed in various phenomenons over the network such as handover, mobility management, and D2D communications [137, 138]. Therefore, a strong and efficient mutual authentication mechanism is required which maintains the freshness of keys and session while securing the applications in the 5G setup.
5. *Secure spectrum-sharing and network slicing*: Network slicing is a key concept of resource allocation and management. The interference in the network imposes signal distortion and noise. To reduce such kinds of issues, optimized and secure spectrum-sharing should be adopted to enable large applications with different user priorities. Such a requirement can be attained through SDN-NFV technologies while leveraging the properties of network slicing [139].

8 Conclusions

This paper provided a detailed description of the security for the 5G networks. The evolution of osmotic and catalytic computing and the details of their implementation and utilization are also presented in this article. The roadmap of different kinds of attacks and their possible solutions are highlighted in the initial parts of this article. Furthermore, the taxonomy on the basis of security requirements is presented, which also includes the comparison of the existing state-of-the-art solutions. In addition, existing surveys, open issues, and security challenges are discussed to provide a research direction. Moreover, this article also provides a security model, "CATMOSIS," which idealizes the incorporation of security features on the basis of catalytic and osmotic computing in the 5G networks.

References

1. E. Hossain, M. Rasti, H. Tabassum, A. Abdelnasser, Evolution toward 5G multi-tier cellular wireless networks: an interference management perspective. IEEE Wirel. Commun. **21**(3), 118–127 (2014)
2. X. Ge, H. Cheng, M. Guizani, T. Han, 5G wireless backhaul networks: challenges and research advances. IEEE Netw. **28**(6), 6–11 (2014)
3. V. Sharma, J.D. Lim, J.N. Kim, I. You, SACA: Self-Aware Communication Architecture for IoT Using Mobile Fog Servers. Mobile Information Systems. 2017; (2017)
4. P. Rost, C.J. Bernardos, A. De Domenico, M. Di Girolamo, M. Lalam, A. Maeder, D. Sabella, D. Wübben, Cloud technologies for flexible 5G radio access networks. IEEE Commun. Mag. **52**(5), 68–76 (2014)
5. Z. Ding, Y. Liu, J. Choi, Q. Sun, M. Elkashlan, H.V. Poor, Application of non-orthogonal multiple access in LTE and 5G networks. arXiv preprint arXiv:1511.08610 (2015)
6. V. Sharma, R. Sabatini, S. Ramasamy, UAVs assisted delay optimization in heterogeneous wireless networks. IEEE Commun. Lett. **20**(12), 2526–2529 (2016)
7. V. Sharma, R. Kumar, P.S. Rana, Self-healing neural model for stabilization against failures over networked UAVs. IEEE Commun. Lett. **19**(11), 2013–2016 (2015)
8. Z. Zhang, X. Chai, K. Long, A.V. Vasilakos, L. Hanzo, Full duplex techniques for 5G networks: self-interference cancellation, protocol design, and relay selection. IEEE Commun. Mag. **53**(5), 128–137 (2015)
9. A.I. Sulyman, A.T. Nassar, M.K. Samimi, G.R. MacCartney, T.S. Rappaport, A. Alsanie, Radio propagation path loss models for 5G cellular networks in the 28 GHz and 38 GHz millimeter-wave bands. IEEE Commun. Mag. **52**(9), 78–86 (2014)
10. D. Shin, V. Sharma, J. Kim, S. Kwon, I. You, Secure and efficient protocol for route optimization in PMIPv6-based smart home IoT networks. IEEE Access. **5**, 11100–11117 (2017)
11. H. Elshaer, F. Boccardi, M. Dohler, R. Irmer, Downlink and uplink decoupling: a disruptive architectural design for 5G networks. In Global Communications Conference (GLOBE-COM), 2014 IEEE 2014 Dec 8 (pp. 1798–1803). IEEE
12. V. Sharma, G. Choudhary, I. You, J.D. Lim, J.N. Kim, Self-enforcing game theory-based resource allocation for LoRaWAN assisted public safety communications. arXiv preprint arXiv:1804.07204 (2018)
13. T.D. Perera, D.N. Jayakody, S. Chatzinotas, V. Sharma, Wireless information and power transfer: issues, advances, and challenges. In Vehicular Technology Conference (VTC-Fall), 2017 IEEE 86th 2017 Sep 24 (pp. 1–7). IEEE
14. M.N. Tehrani, M. Uysal, H. Yanikomeroglu, Device-to-device communication in 5G cellular networks: challenges, solutions, and future directions. IEEE Commun. Mag. **52**(5), 86–92 (2014)
15. X. Duan, X. Wang, Authentication handover and privacy protection in 5G hetnets using software-defined networking. IEEE Commun. Mag. **53**(4), 28–35 (2015)
16. V. Sharma, K. Lee, S. Kwon, J. Kim, H. Park, K. Yim, S.Y. Lee, A consensus framework for reliability and mitigation of zero-day attacks in IoT. Security and Communication Networks. 2017; (2017)
17. V. Sharma, I. You, G. Kul, Socializing drones for inter-service operability in ultra-dense wireless networks using Blockchain. In Proceedings of the 2017 International Workshop on Managing Insider Security Threats 2017 Oct 30 (pp. 81–84). ACM
18. M. Chen, Y. Qian, S. Mao, W. Tang, X. Yang, Software-defined mobile networks security. Mobile Netw. Appl. **21**(5), 729–743 (2016)
19. S. Sun, M. Kadoch, L. Gong, B. Rong, Integrating network function virtualization with SDR and SDN for 4G/5G networks. IEEE Netw. **29**(3), 54–59 (2015)

20. K. Xiao, L. Gong, M. Kadoch, Opportunistic multicast NOMA with security concerns in a 5G massive MIMO system. IEEE Commun. Mag. **56**(3), 91–95 (2018)
21. Y. Wu, A. Khisti, C. Xiao, G. Caire, K.K. Wong, X. Gao, A survey of physical layer security techniques for 5G wireless networks and challenges ahead. IEEE J. Sel. Areas Commun. **36**, 679–695 (2018)
22. V. Sharma, I. You, R. Kumar, ISMA: Intelligent sensing model for anomalies detection in cross platform OSNs with a case study on IoT. IEEE Access **5**, 3284–3301 (2017)
23. M. Condoluci, M.A. Lema, T. Mahmoodi, M. Dohler, 5g IoT industry verticals and network requirements. In Powering the Internet of Things With 5G Networks 2018 (pp. 148–175). IGI Global
24. V. Sharma, I. You, D.N. Jayakody, M. Atiquzzaman, Cooperative trust relaying and privacy preservation via edge-crowdsourcing in social Internet of Things. Future Generation Computer Systems. 2017 Dec 28
25. M. Alenezi, K. Almustafa, M. Hussein, On virtualization and security-awareness performance analysis in 5G cellular networks. J. Eng. Sci. Technol. Rev. **11**(1), 1–9 (2018)
26. D. Schinianakis, Alternative security options in the 5G and IoT era. IEEE Circuits Syst. Mag. **17**(4), 6–28 (2017)
27. Y.H. Lee, A.S. Wang, Y.D. Liao, T.W. Lin, Y.J. Chi, C.C. Wong, N. Shinohara, Q. Yuan, Q. Chen, Wireless power IoT system using polarization switch antenna as polling protocol for 5G mobile network. In Wireless Power Transfer Conference (WPTC), 2017 IEEE 2017 May 10 (pp. 1–3). IEEE
28. A. Costanzo, D. Masotti, Energizing 5G: near-and far-field wireless energy and data trantransfer as an enabling technology for the 5G IoT. IEEE Microw. Mag. **18**(3), 125–136 (2017)
29. A. Rajaram, D.N. Jayakody, K. Srinivasan, B. Chen, V. Sharma, Opportunistic-harvesting: RF wireless power transfer scheme for multiple access relays system. IEEE Access. **5**, 16084–16099 (2017)
30. J. Guan, V. Sharma, I. You, M. Atiquzzaman, Extension of MIH to support FPMIPv6 for optimized heterogeneous handover. arXiv preprint arXiv:1705.09835 (2017)
31. M. Villari, M. Fazio, S. Dustdar, O. Rana, R. Ranjan, Osmotic computing: a new paradigm for edge/cloud integration. IEEE Cloud Comput. **3**(6), 76–83 (2016)
32. V. Sharma, K. Srinivasan, D.N. Jayakody, O. Rana, R. Kumar, Managing service-heterogeneity using osmotic computing. arXiv preprint arXiv:1704.04213 (2017)
33. Maksimović M. The Role of Osmotic Computing in Internet of Things.Infoteh-Jahorina.2018 March
34. V. Sharma, I. You, R. Kumar, P. Kim, Computational offloading for efficient trust management in pervasive online social networks using osmotic computing. IEEE Access. **5**, 5084–5103 (2017)
35. M. Nardelli, S. Nastic, S. Dustdar, M. Villari, R. Ranjan, Osmotic flow: Osmotic computing+ IoT workflow. IEEE Cloud Comput. **4**(2), 68–75 (2017)
36. A. Morshed, P.P. Jayaraman, T. Sellis, D. Georgakopoulos, M. Villari, R. Ranjan, Deep osmosis: holistic distributed deep learning in osmotic computing. IEEE Cloud Comput. **4**(6), 22–32 (2018)
37. T. Rausch, S. Dustdar, R. Ranjan, Osmotic message-oriented middleware for the internet of things. IEEE Cloud Comput. **5**(2), 17–25 (2018)
38. A. Buzachis, A. Galletta, L. Carnevale, A. Celesti, M. Fazio, M. Villari, Towards osmotic computing: analyzing overlay network solutions to optimize the deployment of container-based microservices in fog, edge and IoT environments. In Fog and Edge Computing (ICFEC), 2018 IEEE 2nd International Conference on 2018 May 1 (pp. 1–10). IEEE
39. V. Sharma, I. You, R. Kumar, Resource-based mobility management for video users in 5G using catalytic computing. Comput. Commun. **118**, 120–139 (2018)

40. G. Chopra, R.K. Jha, S. Jain, A survey on ultra-dense network and emerging technologies: security challenges and possible solutions. J. Netw. Comput. Appl. **95**, 54–78 (2017)
41. M.A. Ferrag, L. Maglaras, A. Argyriou, D. Kosmanos, H. Janicke, Security for 4G and 5G cellular networks: a survey of existing authentication and privacy-preserving schemes. J. Netw. Comput. Appl. **101**, 55–82 (2017)
42. D. Fang, Y. Qian, R.Q. Hu, Security for 5G mobile wireless networks. IEEE Access. **6**, 4850–4874 (2018)
43. H. Lin, Z. Yan, Y. Chen, L. Zhang, A survey on network security-related data collection technologies. IEEE Access. **6**, 18345–18365 (2018)
44. D. Rupprecht, A. Dabrowski, T. Holz, E. Weippl, C. Pöpper, On security research towards future mobile network generations. IEEE Commun. Surv. Tutorials **20**, 2518–2542 (2018)
45. I. Ahmad, T. Kumar, M. Liyanage, J. Okwuibe, M. Ylianttila, A. Gurtov, Overview of 5G security challenges and solutions. IEEE Commun. Standards Mag. **2**(1), 36–43 (2018)
46. Y. Gao, S. Hu, W. Tang, Y. Li, Y. Sun, D. Huang, S. Cheng, X. Li, Physical layer security in 5G based large scale social networks: opportunities and challenges. IEEE Access **6**, 26350–26357 (2018)
47. X. Ji, K. Huang, L. Jin, H. Tang, C. Liu, Z. Zhong, W. You, X. Xu, H. Zhao, J. Wu, M. Yi, Overview of 5G security technology. SCIENCE CHINA Inf. Sci. **61**(8), 081301 (2018)
48. N. Panwar, S. Sharma, A.K. Singh, A survey on 5G: the next generation of mobile communication. Phys. Commun. **18**, 64–84 (2016)
49. N.T. Le, M.A. Hossain, A. Islam, D.Y. Kim, Y.J. Choi, Y.M. Jang. Survey of promising technologies for 5G networks. Mobile information systems. 2016
50. R. Yu, J. Ding, X. Huang, M.T. Zhou, S. Gjessing, Y. Zhang, Optimal resource sharing in 5G-enabled vehicular networks: a matrix game approach. IEEE Trans. Veh. Technol. **65**(10), 7844–7856 (2016)
51. N.C. Luong, P. Wang, D. Niyato, Y.C. Liang, F. Hou, Z. Han. Applications of economic and pricing models for resource management in 5G wireless networks: a survey. arXiv preprint arXiv:1710.04771 (2017)
52. H. Zhang, H. Xing, J. Cheng, A. Nallanathan, V.C. Leung, Secure resource allocation for OFDMA two-way relay wireless sensor networks without and with cooperative jamming. IEEE Trans. Industr. Inf. **12**(5), 1714–1725 (2016)
53. M.R. Abedi, N. Mokari, M.R. Javan, H. Yanikomeroglu, Limited rate feedback scheme for resource allocation in secure relay-assisted OFDMA networks. IEEE Trans. Wirel. Commun. **15**(4), 2604–2618 (2016)
54. A.M. Akhtar, X. Wang, L. Hanzo, Synergistic spectrum sharing in 5G HetNets: a harmonized SDN-enabled approach. IEEE Commun. Mag. **54**(1), 40–47 (2016)
55. X. Liu, Y. Liu, X. Wang, H. Lin, Highly efficient 3-D resource allocation techniques in 5G for NOMA-enabled massive MIMO and relaying systems. IEEE J. Sel. Areas Commun. **35**(12), 2785–2797 (2017)
56. H. Yang, B.C. Seet, S.F. Hasan, P.H. Chong, M.Y. Chung, Radio resource allocation for D2D-enabled massive machine communication in the 5G era. In Dependable, Autonomic and Secure Computing, 14th Intl Conf on Pervasive Intelligence and Computing, 2nd Intl Conf on Big Data Intelligence and Computing and Cyber Science and Technology Congress (DASC/PiCom/DataCom/CyberSciTech), 2016 IEEE 14th Intl C 2016 Aug 8 (pp. 55–60). IEEE
57. H. Zhang, N. Liu, X. Chu, K. Long, A.H. Aghvami, V.C. Leung, Network slicing based 5G and future mobile networks: mobility, resource management, and challenges. IEEE Commun. Mag. **55**(8), 138–145 (2017)
58. K. Samdanis, X. Costa-Perez, V. Sciancalepore, From network sharing to multi-tenancy: the 5G network slice broker. IEEE Commun. Mag. **54**(7), 32–39 (2016)
59. B. Wang, K. Huang, X. Xu, L. Jin, Z. Zhong, Y. Wang, Resource allocation for secure communication in $ K $-tier heterogeneous cellular networks: a spatial-temporal perspective. IEEE Access. **6**, 772–782 (2018)

60. J. Li, N. Zhang, Q. Ye, W. Shi, W. Zhuang, X. Shen, Joint resource allocation and online virtual network embedding for 5G networks. In GLOBECOM 2017–2017 IEEE Global Communications Conference 2017 Dec 4 (pp. 1–6). IEEE
61. I. Alqerm, B. Shihada, Sophisticated online learning scheme for green resource allocation in 5g heterogeneous cloud radio access networks. IEEE Trans. Mob. Comput. **17**, 2423–2437 (2018)
62. H. Zhang, N. Yang, K. Long, M. Pan, G.K. Karagiannidis, V.C. Leung, Secure communications in NOMA system: Subcarrier assignment and power allocation. IEEE J. Sel. Areas Commun. **36**(7), 1441–1452 (2018)
63. I.P. Belikaidis, A. Georgakopoulos, E. Kosmatos, V. Frascolla, P. Demestichas, Management of 3.5-GHz spectrum in 5G dense networks: a hierarchical radio resource management scheme. IEEE Veh. Technol. Mag. **13**(2), 57–64 (2018)
64. A. Martin, J. Egaña, J. Flórez, J. Montalbán, I.G. Olaizola, M. Quartulli, R. Viola, M. Zorrilla, Network resource allocation system for QoE-aware delivery of media services in 5G networks. IEEE Trans. Broadcast. **64**(2), 561–574 (2018)
65. Y. He, F.R. Yu, N. Zhao, H. Yin, Secure social networks in 5G systems with mobile edge computing, caching, and device-to-device communications. IEEE Wirel. Commun. **25**(3), 103–109 (2018)
66. L. Tan, *Resource Allocation and Performance Optimization in Communication Networks and the Internet* (CRC Press, Boca Raton, 2017)
67. M. Jiang, M. Condoluci, T. Mahmoodi, Network slicing management & prioritization in 5G mobile systems. In European wireless 2016 May 18 (pp. 1–6)
68. AlQerm I, Shihada B. Enhanced machine learning scheme for energy efficient resource allocation in 5G heterogeneous cloud radio access networks. In IEEE Symposium on Personal, Indoor and Mobile Radio Communications (PIMRC) 2017 Oct 8 (pp. 1–7)
69. X. Duan, *Software-defined Networking enabled Resource Management and Security Provisioning in 5G Heterogeneous Networks*. Electronic Thesis and Dissertation Repository. 4666. https://ir.lib.uwo.ca/etd/4666 [Last Accessed November 2018]
70. Y. Zhu, Efficient resource allocation for 5G hybrid wireless networks (Doctoral dissertation, UCL (University College London))
71. V. Sharma, I. You, F.Y. Leu, M. Atiquzzaman, Secure and efficient protocol for fast handover in 5G mobile Xhaul networks. J. Netw. Comput. Appl. **102**, 38–57 (2018)
72. I. You, J.H. Lee, SPFP: Ticket-based secure handover for fast proxy mobile IPv6 in 5G networks. Comput. Netw. **129**, 363–372 (2017)
73. C.I. Fan, J.J. Huang, M.Z. Zhong, R.H. Hsu, W.T. Chen, J. Lee, ReHand: secure region-based fast handover with user anonymity for small cell networks in 5G. arXiv preprint arXiv:1806.03406 (2018)
74. S. Chen, F. Qin, B. Hu, X. Li, Z. Chen, User-centric ultra-dense networks for 5G: challenges, methodologies, and directions. IEEE Wirel. Commun. **23**(2), 78–85 (2016)
75. K. Munir, E. Zahoor, R. Rahim, X. Lagrange, J.H. Lee, Secure and fault-tolerant distributed location management for Intelligent 5G wireless networks. IEEE Access. **6**, 18117–18127 (2018)
76. V. Sharma, I. You, F. Palmieri, D.N. Jayakody, J. Li, Secure and energy-efficient handover in fog networks using blockchain-based DMM. IEEE Commun. Mag. **56**(5), 22–31 (2018)
77. A. Prasad, P. Lundén, M. Moisio, M.A. Uusitalo, Z. Li, Efficient mobility and traffic management for delay tolerant cloud data in 5G networks. In PIMRC 2015 Aug 30 (pp. 1740–1745)
78. R. Kantola, J. Llorente Santos, N. Beijar, Policy-based communications for 5G mobile with customer edge switching. Secur. Commun. Netw. **9**(16), 3070–3082 (2016)
79. M. Kantor, T. Engel, G. Ormazabal, A policy-based per-flow mobility management system design. In Proceedings of the Principles, Systems and Applications on IP Telecommunications 2015 Oct 6 (pp. 35–42). ACM

80. A. Compagno, X. Zeng, L. Muscariello, G. Carofiglio, J. Auge, Secure producer mobility in information-centric network. In Proceedings of the 4th ACM Conference on Information-Centric Networking 2017 Sep 26 (pp. 163–169). ACM
81. J. Guan, Z. Wei, I. You, GRBC-based network security functions placement scheme in SDS for 5G security. J. Netw. Comput. Appl. **114**, 48–56 (2018)
82. Y. Jung, E. Festijo, M. Peradilla, Joint operation of routing control and group key management for 5G ad hoc D2D networks. In Privacy and Security in Mobile Systems (PRISMS), 2014 International Conference on 2014 May 11 (pp. 1–8). IEEE
83. M. Wang, Z. Yan, Security in D2D communications: a review. In Trustcom/BigDataSE/ISPA, 2015 IEEE 2015 Aug 20 (Vol. 1, pp. 1199–1204). IEEE
84. V.K. Choyi, A. Abdel-Hamid, Y. Shah, S. Ferdi, A. Brusilovsky, Network slice selection, assignment and routing within 5G Networks. In Standards for Communications and Networking (CSCN), 2016 IEEE Conference on 2016 Oct 31 (pp. 1–7). IEEE
85. I.F. Naqvi, A.K. Siddiqui, A. Farooq, IPv6 adoption rate and performance in the 5G wireless internets. In Region 10 Conference (TENCON), 2016 IEEE 2016 Nov 22 (pp. 3850–3858). IEEE
86. M. Schmittner, A. Asadi, M. Hollick, SEMUD: secure multi-hop device-to-device communication for 5G public safety networks. In IFIP Networking Conference (IFIP Networking) and Workshops, 2017 2017 Jun 12 (pp. 1–9). IEEE
87. Y. Zhao, G. Li, W. Qu, A novel cluster-based ultra-dense network technique for 5G and its security issues. In Dependable, Autonomic and Secure Computing, 15th Intl Conf on Pervasive Intelligence & Computing, 3rd Intl Conf on Big Data Intelligence and Computing and Cyber Science and Technology Congress (DASC/PiCom/DataCom/CyberSciTech), 2017 IEEE 15th Intl 2017 Nov 6 (pp. 362–367). IEEE
88. S. Othmen, F. Zarai, A. Belghith, L. Kamoun, Secure Routing Protocol based on Weil Paring for Multi-hop Cellular Network (SRP-MCN). IJCSNS **16**(6), 117 (2016)
89. G. Gomez, F.J. Martin-Vega, F.J. Lopez-Martinez, Y. Liu, M. Elkashlan, Uplink NOMA in large-scale systems: coverage and physical layer security. arXiv preprint arXiv:1709.04693 (2017)
90. M. Forouzesh, P. Azmi, N. Mokari, K.K. Wong, Robust physical layer security for power domain non-orthogonal multiple access-based HetNets and HUDNs, SIC avoidance at eavesdroppers. arXiv preprint arXiv:1806.02013 (2018)
91. D. Kapetanovic, G. Zheng, F. Rusek, Physical layer security for massive MIMO: an overview on passive eavesdropping and active attacks. IEEE Commun. Mag. **53**(6), 21–27 (2015)
92. H.M. Wang, T.X. Zheng, J. Yuan, D. Towsley, M.H. Lee, Physical layer security in heterogeneous cellular networks. IEEE Trans. Commun. **64**(3), 1204–1219 (2016)
93. Z. Qin, Y. Liu, Z. Ding, Y. Gao, M. Elkashlan, Physical layer security for 5G non-orthogonal multiple access in large-scale networks. In Communications (ICC), 2016 IEEE International Conference on 2016 May 22 (pp. 1–6). IEEE
94. Y. Liu, Z. Qin, M. Elkashlan, Y. Gao, L. Hanzo, Enhancing the physical layer security of non-orthogonal multiple access in large-scale networks. IEEE Trans. Wirel. Commun. **16**(3), 1656–1672 (2017)
95. C. Zhang, J. Ge, Z. Xia, H. Du, Graph theory based cooperative transmission for physical-layer security in 5G large-scale wireless relay networks. IEEE Access **5**, 21640–21649 (2017)
96. J. Chen, L. Yang, M.S. Alouini, Physical layer security for cooperative NOMA systems. IEEE Trans. Veh. Technol. **67**(5), 4645–4649 (2018)
97. F. Pan, Y. Jiang, H. Wen, R. Liao, A. Xu, Physical layer security assisted 5G network security. In Vehicular Technology Conference (VTC-Fall), 2017 IEEE 86th 2017 Sep 24 (pp. 1–5). IEEE
98. F. Zhu, F. Gao, H. Lin, S. Jin, J. Zhao, G. Qian, Robust beamforming for physical layer security in BDMA massive MIMO. IEEE J. Sel. Areas Commun. **36**, 775–787 (2018)
99. W.Q. Wang, Z. Zheng, Hybrid MIMO and phased-Array directional modulation for physical layer security in mmWave wireless communications. IEEE J. Sel Areas Commun. **36**(7), 1383–1396 (2018)

100. A. Morgado, K.M. Huq, S. Mumtaz, J. Rodriguez, A survey of 5G technologies: regulatory, standardization and industrial perspectives. Digital Commun. Netw. **4**(2), 87–97 (2018)
101. J. Cheng, W. Chen, F. Tao, C.L. Lin, Industrial IoT in 5G environment towards smart manufacturing. J. Ind. Inf. Integr. **10**, 10–19 (2018)
102. I. Mavromatis, A. Tassi, G. Rigazzi, R.J. Piechocki, A. Nix, Multi-radio 5G architecture for connected and autonomous vehicles: application and design insights. arXiv preprint arXiv:1801.09510 (2018)
103. P.K. Agyapong, M. Iwamura, D. Staehle, W. Kiess, A. Benjebbour, Design considerations for a 5G network architecture. IEEE Commun. Mag. **52**(11), 65–75 (2014)
104. Q. Han, S. Liang, H. Zhang, Mobile cloud sensing, big data, and 5G networks make an intelligent and smart world. IEEE Netw. **29**(2), 40–45 (2015)
105. M.A. Lema, A. Laya, T. Mahmoodi, M. Cuevas, J. Sachs, J. Markendahl, M. Dohler, Business case and technology analysis for 5g low latency applications. IEEE Access **5**, 5917–5935 (2017)
106. F.B. Saghezchi, G. Mantas, J. Ribeiro, M. Al-Rawi, S. Mumtaz, J. Rodriguez, Towards a secure network architecture for smart grids in 5G era. In Wireless Communications and Mobile Computing Conference (IWCMC), 2017 13th International 2017 Jun 26 (pp. 121–126). IEEE
107. G. Arfaoui, P. Bisson, R. Blom, R. Borgaonkar, H. Englund, E. Félix, F. Klaedtke, P.K. Nakarmi, M. Näslund, P. O'Hanlon, J. Papay, A security architecture for 5G networks. IEEE Access **6**, 22466–22479 (2018)
108. J. Santos, T. Wauters, B. Volckaert, F. DeTurck, Fog computing: Enabling the management and orchestration of smart city applications in 5g networks. Entropy **20**(1), 4 (2017)
109. A.H. Celdrán, M.G. Pérez, F.J. Clemente, G.M. Pérez, Towards the autonomous provision of self-protection capabilities in 5G networks. J. Ambient. Intell. Humaniz. Comput. **2018**, 1–4
110. J. Santos, P. Leroux, T. Wauters, B. Volckaert, F. De Turck, Anomaly detection for Smart City applications over 5G low power wide area networks. In NOMS 2018–2018 IEEE/IFIP Network Operations and Management Symposium 2018 Apr 23. IEEE
111. M. Lichtman, R.M. Rao, V. Marojevic, J.H. Reed, R.P. Jover, 5G NR jamming, spoofing, and sniffing: threat assessment and mitigation. arXiv preprint arXiv:1803.03845 (2018)
112. P. Schneider, G. Horn, Towards 5G security. In Trustcom/BigDataSE/ISPA, 2015 IEEE 2015 Aug 20 (Vol. 1, pp. 1165–1170). IEEE
113. S. Luo, J. Wu, J. Li, L. Guo, B. Pei, Toward vulnerability assessment for 5G mobile communication networks. In Smart City/SocialCom/SustainCom (SmartCity), 2015 IEEE International Conference on 2015 Dec 19 (pp. 72–76). IEEE
114. A. Mohseni-Ejiyeh, M. Ashouri-Talouki, SeVR+: Secure and privacy-aware cloud-assisted video reporting service for 5G vehicular networks. In Electrical Engineering (ICEE), 2017 Iranian Conference on 2017 May 2 (pp. 2159–2164). IEEE
115. C. Bouras, A. Kollia, A. Papazois, Teaching network security in mobile 5G using ONOS SDN controller. In Ubiquitous and Future Networks (ICUFN), 2017 Ninth International Conference on 2017 Jul 4 (pp. 465–470). IEEE
116. G. Mantas, N. Komninos, J. Rodriuez, E. Logota, H. Marques, in *Fundamentals of 5G mobile networks*, ed. J. Rodriguez. Security for 5G communications (John Wiley & Sons, Ltd.), p. 207–220. ISBN 9781118867464
117. M. Svensson, N. Paladi, R. Giustolisi, *5G: Towards Secure Ubiquitous Connectivity Beyond 2020, Kista* (Sweden: Swedish Institute of Computer Science, 2015), p. 24. SICS Technical Report, ISSN 1100-3154; 2015:08.
118. A. Gupta, R.K. Jha, S. Jain, Attack modeling and intrusion detection system for 5G wireless communication network. Int. J. Commun. Syst. **30**(10), e3237 (2017)
119. I. Ahmad, T. Kumar, M. Liyanage, J. Okwuibe, M. Ylianttila, A. Gurtov, 5G security: analysis of threats and solutions. In Standards for Communications and Networking (CSCN), 2017 IEEE Conference on 2017 Sep 18 (pp. 193–199). IEEE
120. N.G. Alliance. *5G Security Recommendations Package 2: Network Slicing*. https://www.ngmn.org/fileadmin/user_upload/160429_NGMN_5G_Security_Network_Slicing_v1_0.pdf. Page (1–12) Version 1.0, 27 –April-2016 [Last Accessed November 2018]

121. C. Felita, M. Suryanegara, 5G key technologies: identifying innovation opportunity. In QiR (Quality in Research), 2013 International Conference on 2013 Jun 25 (pp. 235–238). IEEE
122. N. Yang, L. Wang, G. Geraci, M. Elkashlan, J. Yuan, M. Di Renzo, Safeguarding 5G wireless communication networks using physical layer security. IEEE Commun. Mag. **53**(4), 20–27 (2015)
123. L. Sun, Q. Du, Physical layer security with its applications in 5G networks: a review. China Commun. **14**(12), 1–4 (2017)
124. K. Gai, M. Qiu, L. Tao, Y. Zhu, Intrusion detection techniques for mobile cloud computing in heterogeneous 5G. Secur. Commun Netw. **9**(16), 3049–3058 (2016)
125. R. Atat, L. Liu, H. Chen, J. Wu, H. Li, Y. Yi, Enabling cyber-physical communication in 5g cellular networks: challenges, spatial spectrum sensing, and cyber-security. IET Cyber-Phys. Syst. Theor. Appl. **2**(1), 49–54 (2017)
126. O. Mämmelä, J. Hiltunen, J. Suomalainen, K. Ahola, P. Mannersalo, J. Vehkaperä, Towards micro-segmentation in 5G network security. In European Conference on Networks and Communications (EuCNC 2016) Workshop on Network Management, Quality of Service and Security for 5G Networks 2016 Jun
127. P. Gandotra, R.K. Jha, A survey on green communication and security challenges in 5G wireless communication networks. J. Netw. Comput. Appl. **96**, 39–61 (2017)
128. A. Dorri, S.S. Kanhere, R. Jurdak, P. Gauravaram, Blockchain for IoT security and privacy: the case study of a smart home. In Pervasive Computing and Communications Workshops (PerCom Workshops), 2017 IEEE International Conference on 2017 Mar 13 (pp. 618–623). IEEE
129. T. Gupta, G. Choudhary, V. Sharma, A survey on the security of Pervasive Online Social Networks (POSNs). arXiv preprint arXiv:1806.07526 (2018)
130. J.A. Oravec, Emerging "cyber hygiene" practices for the Internet of Things (IoT): professional issues in consulting clients and educating users on IoT privacy and security. In Professional Communication Conference (ProComm), 2017 IEEE International 2017 Jul 23 (pp. 1–5). IEEE
131. A. Dorri, S.S. Kanhere, R. Jurdak, P. Gauravaram, LSB: a lightweight scalable BlockChain for IoT security and privacy. arXiv preprint arXiv:1712.02969 (2017)
132. V. Sharma, R. Kumar, W.H. Cheng, M. Atiquzzaman, K. Srinivasan, A. Zomaya, NHAD: Neuro-fuzzy based Horizontal Anomaly Detection in online social networks. IEEE Trans. Knowl. Data Eng. **30**, 2171–2184 (2018)
133. V. Sharma, J. Kum, S. Kwon, I. You, F.-Y. Leu, An overview of 802.21a-2012 and its incorporation into IoT-Fog networks using osmotic framework. In IoTaaS 2017 – 3rd EAI International Conference on IoT as a Service, vol. 3, pp. 1–6. EAI, Sept. 2017
134. I. You, S. Kwon, G. Choudhary, V. Sharma, J.T. Seo, An enhanced LoRaWAN security protocol for privacy preservation in IoT with a case study on a smart factory-enabled parking system. Sensors (Basel, Switzerland) **18**(6), 1888 (2018)
135. M.A. Khan, K. Salah, IoT security: review, blockchain solutions, and open challenges. Futur. Gener. Comput. Syst. **82**, 395–411 (2018)
136. B. Baranidharan, Internet of Things (IoT) technologies, architecture, protocols, security, and applications: a survey, in *In Handbook of Research on Cloud and Fog Computing Infrastructures for Data Science*, (IGI Global, 2018), pp. 149–174
137. V. Sharma, J. Kim, S. Kwon, I. You, K. Lee, K. Yim, A framework for mitigating zero-day attacks in IoT. arXiv preprint arXiv:1804.05549 (2018) In: Conference on Information Security and Cryptography (CISC-S'17) (South Korea, 2017), pp. 1–6
138. V. Sharma, J. Kum, S. Kwon, I. You, F.-Y. Leu, Fuzzy-based protocol for secure remote diagnosis of IoT devices in 5g networks. In IoTaaS 2017 – 3rd EAI International Conference on IoT as a Service, vol. 3, pp. 1–6. EAI, Taiwan, Sept. 2017
139. J. Ni, X. Lin, X.S. Shen, Efficient and secure service-oriented authentication supporting network slicing for 5G-enabled IoT. IEEE J. Sel. Areas Commun. **36**(3), 644–657 (2018)

Gaurav Choudhary received his B.Tech. degree in computer science and engineering from Rajasthan Technical University in 2014 and master's degree in cyber security from Sardar Patel University of Police in 2017. He is currently pursuing Ph.D. degree in the Department of Information Security Engineering, Soonchunhyang University, Asan, South Korea. His areas of research and interests are UAVs, mobile and internet security, IoT security, network security, and cryptography.

Vishal Sharma received his Ph.D. and B.Tech. degrees in computer science and engineering from Thapar University (2016) and Punjab Technical University (2012), respectively. He worked at Thapar University as a lecturer from April 2016 to October 2016. From November 2016 to September 2017, he was a joint post-doctoral researcher in MobiSec Lab. at Department of Information Security Engineering, Soonchunhyang University, and Soongsil University, Republic of Korea. Dr. Sharma is now a research assistant professor in the Department of Information Security Engineering, Soonchunhyang University, The Republic of Korea. Dr. Sharma received three best paper awards from the IEEE International Conference on Communication, Management and Information Technology (ICCMIT), Warsaw, Poland, in April 2017; from CISC-S'17 South Korea in June 2017; and from IoTaas Taiwan in September 2017. He is the member of IEEE, a professional member of ACM and past Chair for ACM Student Chapter – TU Patiala. He has authored/coauthored more than 70 journal/conference articles and book chapters. He serves as the program committee member for the Journal of Wireless Mobile Networks, Ubiquitous Computing, and Dependable Applications (JoWUA). He was the track chair of MobiSec'16 and AIMS-FSS'16, and PC member and reviewer of MIST'16 and MIST'17, respectively. He was the TPC member of ITNA-CIEEE TCBD'17 and serving as TPC member of ICCMIT'18, CoCoNet'18, and ITNAC-IEEE TCBD'18. Also, he serves as a reviewer for various IEEE Transactions and other journals. His areas of research and interests are 5G networks, UAVs, estimation theory, and artificial intelligence.

Physical Layer Security in 5G Hybrid Heterogeneous Networks

Anum Umer and Syed Ali Hassan

1 Introduction

The paradigm of communication is shifting toward fifth-generation (5G) technology as the increased traffic demands cannot be met with existing conventional sub-6 GHz communication. This is because today's era is saturated with widespread usage of smart devices and ever-increasing wireless data traffic. The 5G communication encompasses certain key enabler technologies at the physical layer that makes 5G networks plausible. This includes the massive multiple-input multiple-output (MIMO) technology and millimeter wave (mmW) communication, at 10 to 300 GHz radio frequency bands with bandwidth above 2 GHz, which are envisioned in conjunction with heterogeneous cellular networks (HetNets) [1, 2]. The HetNets provide enhanced coverage and throughput to the end users by operating in such a fashion that they create a layer of overlay deployment of small cells over the existing sub-6 GHz macro cells, thus, bringing network close to the user. Small cells consist of low-powered base stations (BS) with variable operating frequencies and communication ranges [3].

In HetNets, massive MIMO technology works by deploying a range of large-scale antenna arrays at the transmitting nodes to produce highly directional beam gains and optimal radio spectral efficiency [4]. Whereas mmW communication cells have limited coverage because of operation at smaller wavelength and higher path loss [5] due to sensitivity to blockages and severe propagation losses, however, the small wavelength of mmW communication allows placement of large array of antennas in small area; therefore, beamforming can be implemented to compensate

A. Umer · S. A. Hassan (✉)
School of Electrical Engineering and Computer Science (SEECS), National University of Sciences and Technology (NUST), Islamabad, Pakistan
e-mail: ali.hassan@seecs.edu.pk

D. N. K. Jayakody et al. (eds.), *5G Enabled Secure Wireless Networks*,
https://doi.org/10.1007/978-3-030-03508-2_4

for path losses, additional noise power, and out-of-cell interference [7]. It has been proved through measurements that there are significant differences in path loss of line-of-sight (LoS) and non-line-of-sight (NLoS) mmW propagation paths [2, 6].

Shifting the focus from coverage probability and achievable rate at the end user, in 5G communication, investigation of its secrecy and information integrity aspects has recently become an important discussion in both academia and industry. The subsequent discussion will aim to analyze the aspects of physical layer security (PLS) in massive MIMO-enabled hybrid HetNet with mmW small cells, since the physical layer security (PLS) presents an important, low complexity solution for protection of confidential information in complex networks and the aforementioned network presents a common deployment scenario for future 5G communication networks.

In this chapter, we study the PLS in three-tier hybrid HetNets with massive MIMO in macro tier and mmW frequency and sub-6 GHz small cells. Specifically, we analyze the performance of the proposed network, through analytical modeling and simulation, in terms of secrecy outage probability, secrecy spectral efficiency (SSE), and secrecy energy efficiency (SEE). The developed tractable approach of analysis of the network accounts for key features of mmW communication, massive MIMO technology, beamforming gains, number of transmitting antennas at BSs, and node densities. Secrecy outage probability modeling for each tier of the network is performed to quantify the effects of various system parameters on secrecy outage. Following it, tractable model is developed for achievable ergodic rate at the legitimate users and eavesdropper to model average achievable secrecy rate of the network. The SSE and SEE of massive MIMO-enabled three-tier hybrid HetNet are modeled and studied based on the aforementioned average achievable secrecy rate. Finally, with the help of numerical simulations, analytical models are verified, and relation between secrecy outage probability (SOP), SEE, SSE, key features of mmW communication, massive MIMO technology, beamforming gains, number of transmitting antennas at BSs, and node densities is studied.

2 Background

PLS in 5G communication networks is an emerging field of study in wireless technologies. In this respect, major research studies have been done in recent years. Liang et al. [9] and Wang et al. [10] have shown in their work the effects of fading on secrecy outage probability of the network. Wang et al. [8] and Wang et al. [11] investigated that the secrecy performance can be improved by degrading the eavesdropper channel with the help of techniques such as jamming, artificial noise, beam forming, and Wyner codes. Lv et al. [12] designed the techniques for spectrum allocation and transmit beamforming that can improve secrecy rate in two-tier HetNet. Wu et al. [13] stochastically modeled secrecy outage probability and throughout of a HetNet and studied the model for different system parameters. Wang et al. [14] proposed a mobile association policy, based on access threshold, for K-tier HetNet, and Deng et al. [15] worked on secrecy rate in massive MIMO-enabled HetNets. Wang et al. [16] investigated jamming aspects of secrecy in a network where multiple antenna-aided transmitter transmits to a single-antenna user. Xu et al. [17] studied secrecy in HetNets based on coordinated multipoint scheme, and Wang et al. [18] derived and analyzed energy efficiency and secrecy rate in massive MIMO-based heterogeneous centralized radio access network (C-RAN) and showed that secrecy improves with both centralized and distributed large-scale antenna systems in such networks. All the aforementioned works are focused on PLS in conventional sub-6 GHz networks. Sharma et al. [19] present the novel methodology for secure exchange of 3D way points between unmanned aerial vehicles (UAVs), and Sharma et al. [20] discuss a DMM scheme for secure and energy-efficient handover between the mobile nodes in 5G communication empowered fog networks. Jameel et al. [21] studied the secrecy rate outage probability at the legitimate transmitter and receiver in the presence of eavesdroppers capable of energy harvesting and information decoding.

mmW communication has different propagation properties from sub-6 GHz communication, and investigation of its PLS properties is an emerging study. For instance, Wang et al. [22] studied the secrecy properties of point-to-point mmW link and showed that mmW systems have better secrecy as compared to the conventional systems. Vuppala et al. [23] derived the effects of blockages on secrecy rate of a networks with both sub-6 GHz and mmW frequency cells. Wang et al. [24] analyzed the secrecy outage probability in a mmW communication network with omnidirectional single-antenna-assisted users and eavesdroppers. Gong et al. [25] proposed a beamforming scheme for mmW communication networks to maximize their secrecy rate. This approach has been specifically developed for two-way amplify-and-forward MIMO relaying mmW networks. Umer et al. [26] studied the coverage and rate trends of a network with mmW base stations installed in combination with massive MIMO-enabled hybrid HetNets. Umer et al. [27] used stochastic geometry to model and discuss the secrecy outage probability aspects of aforementioned network model.

The rest of the chapter is organized as follows. We first develop the system model and channel model before proceeding to derive the analytical model for the secrecy outage probability, SEE, and SSE of the network. Subsequently, the numerical results for secrecy outage probability and other parameters are discussed. Finally we draw conclusion at the end of the chapter.

3 The System Layout

Consider a time-division duplex-based downlink transmission scenario of three-tier hybrid HetNet consisting of macro cells operating at sub-6 GHz at tier 1 and small cells operating at sub-6 GHz and mmW band at tier 2 and tier 3, respectively, as shown in Fig. 1. The BSs of each tier, legitimate users and eavesdroppers, are spatially distributed, following a two-dimensional homogeneous Poisson point process (HPPP) with intensity Φ_k and density λ_k where $k \in \{1, 2, 3\}$, Φ_u, and its

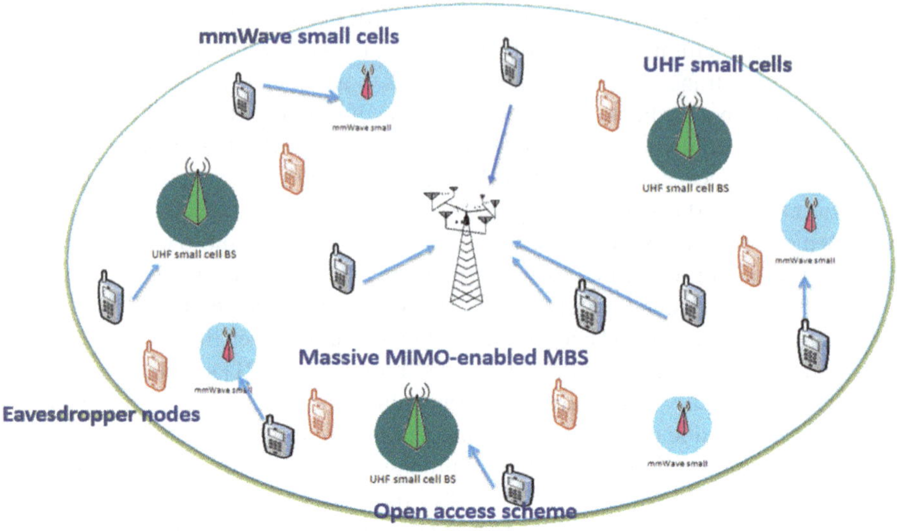

Fig. 1 Proposed three-tier massive MIMO-enabled hybrid HetNet with mmW small cells

density λ_u and Φ_e with density λ_e, respectively. Each BS of kth tier has transmit power P_k and path loss exponent α_k. Thus, the transmission from BS to the legitimate user takes place in the presence of spatially distributed eavesdroppers. The users and eavesdroppers are equipped with single omnidirectional antenna, and massive MIMO is implemented at macro cells where each macro base station (MBS) is equipped with multiple antennas forming an array, i.e., each MBS has N antennas that simultaneously transmit to U users ($N \gg U \geq 1$) with equal power distribution among all users [28]. Assume that channel state information (CSI) is known at the MBS that uses zero-forcing beamforming (ZFBF) to transmit U data streams [29]. Following Slivnyak's theorem, we perform the analysis of the network for a typical user located at the origin. A set of parameters used in this paper are outlined in Table 1.

Since mmW frequencies suffer significant path loss, therefore, directional beamforming is implemented at the mmW small cells BSs. We employ a sectored model to define the antenna pattern at the BSs of mmW tier such that constant array gains are assumed for the main lobes as well as the side lobes. Beam direction is uniformly distributed between $(0, 2\pi]$. The effective antenna gain, G_l, at a typical receiver, r, for an interferer, t, for possible directions $l = \{1, 2, 3, 4\}$ is given as

$$G_l = \begin{cases} a_l = M_r M_t & \text{with} \quad \text{prob.} \quad p_l = (\frac{\theta_r}{2\pi}\frac{\theta_t}{2\pi}) \\ a_l = M_r m_t & \text{with} \quad \text{prob.} \quad p_l = (\frac{\theta_r}{2\pi}(1 - \frac{\theta_t}{2\pi})) \\ a_l = m_r M_t & \text{with} \quad \text{prob.} \quad p_l = ((1 - \frac{\theta_r}{2\pi})\frac{\theta_t}{2\pi}) \\ a_l = m_r m_t & \text{with} \quad \text{prob.} \quad p_l = ((1 - \frac{\theta_r}{2\pi})(1 - \frac{\theta_t}{2\pi})), \end{cases} \tag{1}$$

Table 1 Table of nomenclature

Parameter	Description	Parameter	Description
λ_k	Tier k BS density	f_k	Operating frequency of kth tier
B_k	Bandwidth of kth tier	α_k	Path loss exponent of kth tier
λ_u	User density	P_k	kth tier BS transmission power
λ_e	Eavesdropper density	N	Number of antennas at MBS
U	Users served by a MBS	α_L	Patloss exponent of LoS mmW BS
α_N	Patloss exponent of NLoS mmW BS	N_N	Nakagami parameter at NLoS mmW BS
N_L	Nakagami parameter at LoS mmW BS	M_r	Main lobe gain at receiver
M_t	Main lobe gain at transmitter/ interferer	m_r	Side lobe gain at receiver
m_t	Side lobe gain at transmitter/ interferer	θ_r	Beamwidth at receiver
θ_t	Beamwidth at transmitter/ interferer	σ^2	Noise power
R_{Sk}	Average achievable secrecy rate at kth tier	β	Blockage density [30]
R_k	Achievable ergodic rate at user associated with kth tier	$SINR_k^u$	SINR at user associated with kth tier
$SINR_k^{e^*}$	SINR at most malicious eavesdropper associated with kth tier	P_k^{total}	Power consumption in kth tier

where M_j, θ_j, m_j, and p_l donate the main lobe gain, beamwidth, side lobe gain, and the probability of occurrence of certain gain G_l, where $j = \{t, r\}$. It is assumed that the typical transmitting BS, t, and receiver, r, are perfectly aligned; hence maximum directivity gain $M_r M_t$ can be achieved.

Following the legitimate typical user, antenna gain seen at the eavesdropper $e \in \Phi_e$ from serving BS, t, is defined as

$$G_e = \begin{cases} M_e M_t & \text{with prob.} \quad \left(\frac{\theta_e}{2\pi}\frac{\theta_t}{2\pi}\right) \\ M_e m_t & \text{with prob.} \quad \left(\frac{\theta_e}{2\pi}\left(1 - \frac{\theta_t}{2\pi}\right)\right) \\ m_e M_t & \text{with prob.} \quad \left(\left(1 - \frac{\theta_e}{2\pi}\right)\frac{\theta_t}{2\pi}\right) \\ m_e m_t & \text{with prob.} \quad \left(\left(1 - \frac{\theta_e}{2\pi}\right)\left(1 - \frac{\theta_t}{2\pi}\right)\right), \end{cases} \tag{2}$$

where M_e is the main lobe gain, m_e is the back lobe gain, and θ_e is the main lobe beamwidth at the eavesdropper e, respectively.

Analysis is performed for the typical user to whom mmW BS can form line-of-sight (LoS) or non-line-of-sight (NLoS) link. Moreover, we infer that the typical user at the origin will only form the LoS link to the mmW small cell BS when there is no blockage in path of their link. Thus, to define blockage model for mmW communication in this network, in accordance with independent thinning theorem, we divide Φ_3 to Φ_3^L and $\Phi_3^N = \Phi_3/\Phi_3^L$, with densities $p(x)\lambda_3$ and $(1 - p(x))\lambda_3$, using LoS probability function $p(x)$, as independent PPPs of LoS and NLoS mmW small cells, respectively [30]. We define $p(x)$ as a probability measure that a link of length x is LoS, i.e., $1 - p(x)$ is NLoS probability of a link. Based on stochastic blockage models, $p(x)$ is given as $p(x) = e^{-\beta x}$ where β is dependent on statistics of blockages at a certain cell location and x is the measure of link distance from the serving BS to the typical user [31].

It is assumed that the small-scale fading in sub-6 GHz links is independent and identically distributed (i.i.d) and follows a Rayleigh distribution, whereas for mmW links, it is assume to be independent Nakagami fading with Nakagami fading parameter N_L and N_N for LoS and NLoS links, respectively. Nakagami fading parameters are considered as positive integers for ease of tractability [37].

An open access scheme for users' connectivity to BSs is assumed where user connects with any tier BS based on the maximum average received power. Thus, the typical user will connect to tier j if

$$j = \arg \max_{k \in \{1,2,3\}} \overline{P_k} L_k(x), \tag{3}$$

where $\overline{P_k} = \frac{P_k}{(\frac{4\pi}{\lambda_c})^2}$ and $L_k(x) = x_k^{-\alpha_k}$ are the normalized transmission power of the kth tier and path loss function. λ_c is the carrier wavelength and x is the link distance from typical user to serving BS.

We define the average received power at a typical user connected with the MBS tier as

$$P_{r,1} = G_M \frac{\overline{P_1}}{U} L_{j,M}(x), \tag{4}$$

where $G_M = (N - U + 1)$ is the array gain for ZFBF transmission and $L_{j,M}(x) = x_3^{-\alpha_3}$ is the path loss function [29]. Likewise, we define average received power at the typical user associated with small cell tier BS as

$$P_{r,i} = \overline{P_i} G_i L_i(x), \quad \text{where} \quad i \in \{2, 3\}, \tag{5}$$

where G_i is given as

$$G_i = \begin{cases} 1, \text{sub} - 6\,\text{GHz small cell}, \\ G_l, \text{defined in (1)}, \ \text{mmW small cell}, \end{cases}$$

For the channel modeling of considered three-tier HetNet, we model the signal-to-interference-plus-noise ratio (SINR) for the entire network. It is assumed that the transmission channels of eavesdroppers and legitimate user are independent of each other, and distortion of eavesdropper channel with interference leads to enhanced secrecy performance of overall network. We define most malicious eavesdropper for any transmission link as one with the highest received SINR; thus it dominates the secrecy rate. The SINRs discussed below are for the most malicious eavesdropper, i.e., $\mathrm{SINR}_i^{e^*} = \max_{e\in\Phi_e}\{\mathrm{SINR}_i^e\}$ where $i \in \{M, S, m\}$.

We define received SINR for a typical receiver and any eavesdropper connected with the MBS $b_{o,M}$ as

$$\mathrm{SINR}_M^u = \frac{\frac{\overline{P_1}}{U} h_{o,M} L_{o,M}(x)}{\sigma^2 + \sum_{v\in\Phi_1\backslash b_{o,M}} \frac{\overline{P_1}}{U} h_{v,M} L_{v,M}(x_v) + I_S}, \tag{6}$$

$$\mathrm{SINR}_M^{e^*} = \frac{\frac{\overline{P_1}}{U} h_{e,M} r_e^{-\alpha_1}}{\sigma^2 + \sum_{v\in\Phi_1} \frac{\overline{P_1}}{U} h_{v,M} L_{v,M}(x_v) + I_S}, \tag{7}$$

where $L_s(x_s) = x_s^{-\alpha_2}$, $L_{o,M}(x) = x^{-\alpha_1}$, and $I_S = \sum_{s\in\Phi_2} \overline{P_2} h_s L_s(x_s)$ are the intercell interferences from sub-6 GHz small cells and $h_s \sim \exp(1)$, $h_{v,M} \sim \Gamma(U, 1)$, and $h_{o,M} \sim \Gamma(N - U + 1, 1)$ are the small-scale fading gains at the typical user from the interfering channel, interfering MBS, and from serving MBS for U users [29]. Here, $h_{e,M} \sim \exp(1)$ is fading gain at eavesdropper at the distance r_e from the serving BS, x_s and x_v are the distances between the receiver and small cell BS and receiver and MBS v, and σ^2 is the noise power.

We define the SINR of a typical receiver with link distance x and an eavesdropper with link distance r_e connected with the sub-6 GHz band small cell BS $b_{o,S}$ as

$$\mathrm{SINR}_S^u = \frac{\overline{P_2} h_{o,S} L_{o,S}(x)}{\sigma^2 + \sum_{s\in\Phi_2\backslash b_{o,S}} \overline{P_2} h_{s,S} L_{s,S}(x_s) + I_{\overline{M}}}, \tag{8}$$

$$\mathrm{SINR}_S^{e^*} = \frac{\overline{P_2} h_{e,S} r_e^{-\alpha_2}}{\sigma^2 + \sum_{s\in\Phi_2} \overline{P_2} h_{s,S} L_{s,S}(x_s) + I_{\overline{M}}}, \tag{9}$$

where $L_{s,S}(x_s) = x_s^{-\alpha_2}$, $L_{o,S}(x) = x^{-\alpha_2}$, and $I_{\overline{M}} = \sum_{v\in\Phi_1} \frac{\overline{P_1}}{U} h_v L_v(x_v)$ are the intercell interferences from macro cells and $h_{s,S} \sim \exp(1)$, $h_{o,S} \sim \exp(1)$, $h_{e,S} \sim \exp(1)$, and $h_j \sim \Gamma(U, 1)$ are the small-scale fading gains from the interfering channel. Here, x_s and x_v are the link distance between typical user and small cell BS s and MBS v, respectively.

The SINR for the typical receiver and an eavesdropper connected with mmW small cell $b_{o,m}$ is defined as

$$\mathrm{SINR}_m^u = \frac{\overline{P_3} M_r M_t h_{o,m} L_{o,m}(x)}{\sigma^2 + \overline{P_3} \sum_{q\in\{L,N\}} \sum_{l\in\Phi_3^q\backslash b_{o,m}} G_l h_{l,m} L_{l,m}(x_l)}, \tag{10}$$

$$\mathrm{SINR}_m^{e^*} = \frac{\overline{P_3} G_e h_{e,m} r_e^{-\alpha_3^{(q)}}}{\sigma^2 + \overline{P_3} \sum_{q\in\{L,N\}} \sum_{l\in\Phi_3^q} G_l h_{l,m} L_{l,m}(x_l)}, \tag{11}$$

where $L_{o,m}(x) = x^{-\alpha_3^{(q)}}$ is path loss, $h_{o,m}$ and $h_{e,m}$ are small-scale fading gains, and G_l and G_e are the directivity gains of interfering BSs, given by (1) and (2). Here, $q \in \{L, N\}$ identifies the interfering link as either LoS (L) or NLoS (N), respectively.

The power consumption model for the assumed three-tier hybrid HetNet needs to be quantified for the evaluation of SSE and SEE, respectively. We define total power consumption at MBS as

$$P_1^{\text{total}} = \rho_1 + \frac{P_1}{\epsilon_1} + \sum_{t=1}^{3} (U)^t (\Delta_t + N\Lambda_t), \tag{12}$$

where ρ_1 and ϵ_1 are load-independent circuit power of BD and efficiency of power amplifier, respectively. Parameters Δ_t and Λ_t are dependent on the length of transceiver chains and coding, decoding, and precoding involved in the transmission [32].

The power consumption in sub-6 GHz and mmW small cells is defined as

$$P_i^{\text{total}} = \rho_i + \frac{P_i}{\epsilon_i} \quad \text{for} \quad i \in \{2, 3\}, \tag{13}$$

where $i \in \{2, 3\}$, ρ_i is load-independent circuit power, and ϵ_i is the efficiency of power amplifier of the BS of the i-th tier.

To characterize the secure transmission scenario for the proposed network, we first assume that all tier links are eavesdropped such that eavesdroppers do not attempt attacks to change information transmitted from BS to legitimate receiver, i.e., passive non-colluding eavesdropping. A secrecy coding scheme, Wyner code, is adopted at each transmission link for the protection of confidential messages from intrusion from eavesdroppers. Under the Wyner coding scheme, each BS encodes data using this scheme before transmitting it to the legitimate user [33]. Thus, the rate of the transmitted confidential message signal R_m and rate of transmitted code words R_c are defined at the BS before data transmission commences. The cost of maintaining the confidential message secrecy from eavesdroppers is $R_c - R_m$ [33]. It is assumed that the aforementioned rates remain fixed during transmission [34, 35]. During transmission from BS to the legitimate user, whenever the wiretapping capacity of the link from the serving BS to the eavesdropper R_e is higher than the rate $R_c - R_m$, secrecy outage event occurs. Thus, the secrecy outage probability is defined as $P_{so}^k(\gamma_e) = \Pr(\text{SINR}_k^e > \gamma_e)$, i.e., the SINR at any eavesdropper node is higher than threshold. We quantify SSE and SEE based on secrecy outage probability where SSE is the average secrecy rate per unit bandwidth and SEE is the secrecy performance of a three-tier hybrid HetNet based on unit energy consumption.

4 System Performance Evaluation

We specify the average achievable rate associated with successful transmission of confidential information from BS to the legitimate user in proposed hybrid HetNet as secrecy transmission capacity constraint [33, 36]. Thus, the average achievable secrecy rate for the network tiers is represented as

$$R_{Sk} = [R_k - R_k^e]^+ \quad \text{for} \quad k \in \{1, 2, 3\}, \tag{14}$$

where $[y]^+ = \max\{0, y\}$, $R_k = \mathbb{E}[\log_2(1+\text{SINR}_k^u)]$, and $R_k^e = \mathbb{E}[\log_2(1+\text{SINR}_k^{e^*})]$ are the average achievable ergodic rates of the channel between the serving kth tier BS and the typical receiver and most malicious eavesdropper . As we are performing the analysis for the most malicious eavesdropper, therefore, the average achievable ergodic rate cannot exceed R_k^e. The ergodic transmission rate of the serving BS is dependent on the CSI of the link between itself and legitimate user only as CSI of eavesdroppers is unknown at the BS because of their non-colluding nature.

4.1 Achievable Rates

Following up from (14), we derive the achievable ergodic rate at the legitimate user connected with MBS as

$$R_1 = \frac{1}{\ln 2} \int_0^\infty \frac{P_C^1(\gamma)}{1+\gamma} d\gamma, \tag{15}$$

where $P_C^1(\gamma) = \int_0^\infty P_C^1(\gamma, x) f_{X_1}(x) dx$ is the complementary cumulative distribution function (CCDF) of SINR_M^u, $f_{X_1}(x)$ is the probability density function (PDF) of the distance between the MBS and typical receiver, and $P_C^1(\gamma, x)$ is the conditional coverage probability for the given distance x between the typical user and serving MBS [27]. In this scenario when the legitimate receiver is connected with MBS, the average ergodic rate on the link between serving BS and the most malicious eavesdropper is given by

$$R_1^e = \frac{1}{\ln 2} \int_0^\infty \frac{1 - P_{so}^1(\gamma_e)}{1+\gamma_e} d\gamma_e, \tag{16}$$

where $P_{so}^1(\gamma_e)$ is CDF of $\text{SINR}_M^{e^*}$. Following up from (14), we derive the achievable ergodic rate at the legitimate user connected with sub-6 GHz small cell as

$$R_2 = \frac{1}{\ln 2} \int_0^\infty \frac{P_C^2(\gamma)}{1+\gamma} d\gamma, \tag{17}$$

where $P_C^2(\gamma) = \int_0^\infty P_C^2(\gamma, x) f_{X_2}(x) dx$ is the CCDF of SINR_S^u, $f_{X_2}(x)$ is the PDF of the distance between the serving sub-6 GHz small cell BS and typical user, and $P_C^2(\gamma, x)$ is the conditional coverage probability of the typical user.

In a scenario when the legitimate receiver is connected with sub-6 GHz small cell BS, the average ergodic rate on the link between serving BS and the most malicious eavesdropper is given by

$$R_2^e = \frac{1}{\ln 2} \int_0^\infty \frac{1 - P_{so}^2(\gamma_e)}{1 + \gamma_e} d\gamma_e, \tag{18}$$

where $P_{so}^2(\gamma_e)$ is the CDF of $\mathrm{SINR}_S^{e^*}$.

Following up from (14), we derive the achievable ergodic rate at the legitimate user connected with mmW small cell as

$$R_3 = \frac{1}{\ln 2} \int_0^\infty \frac{P_C^3(\gamma)}{1 + \gamma} d\gamma, \tag{19}$$

where $P_C^3(\gamma) = \sum\limits_{q \in \{L,N\}} A_{3,q} P_C^{3,q}(\gamma)$ is the CCDF of SINR_m^u and $P_C^{3,L}(\gamma)$ and $P_C^{3,N}(\gamma)$ are defined as the conditional coverage probability when the typical receiver, connected with mmW small cell, connects with the BS in Φ_3^L and Φ_3^N, respectively. $A_{3,q}$ are the probabilities of typical receiver associating with LoS or NLoS link.

In a scenario when the legitimate receiver is connected with mmW small cell BS, the average ergodic rate on the link between serving BS and the most malicious eavesdropper is given by

$$R_3^e = \frac{1}{\ln 2} \int_0^\infty \frac{1 - P_{so}^3(\gamma_e)}{1 + \gamma_e} d\gamma_e, \tag{20}$$

where $P_{so}^2(\gamma_e)$ is the CDF of $\mathrm{SINR}_m^{e^*}$. By substituting (19) and (20) in (14), we obtain the average achievable secrecy rate for mmW tier.

4.2 Physical Layer Security Parameters

The secrecy outage probability, P_{so}, for the entire network is defined as

$$P_{so} = \sum_{k=1}^{3} P_{so}^k A_k, \tag{21}$$

where A_k is the association probability of tier k. Similarly, a lower bound on the SSE, using the law of total expectation, is given by

$$\mathrm{SSE}^L = \sum_{k=1}^{3} A_k \times \mathrm{SSE}_k, \tag{22}$$

where $\mathrm{SSE}_1 = U \times R_{S1}$ is the value of SSE for massive MIMO-enabled sub-6 GHz macro tier and $\mathrm{SSE}_k = R_{Sk}$ for $k \in \{2, 3\}$ is the value of SSE for sub-6 GHz and mmW small cell tiers, respectively. Average achievable secrecy rate R_{Sk} for each tier k of the network is given by (14).

The lower bound on SEE for the proposed network is defined as [28, 38]:

$$\mathrm{SEE}^L = \sum_{k=1}^{3} A_k \times \mathrm{SEE}_k, \tag{23}$$

where $\mathrm{SEE}_1 = \frac{U \times R_{S1}}{P_1^{\text{total}}}$ is the value of SEE for massive MIMO sub-6 GHz enabled macro tier and $\mathrm{SEE}_k = \frac{R_{Sk}}{P_k^{\text{total}}}$ for $k \in \{2, 3\}$ is SEE for sub-6 GHz and mmW small cell tiers, respectively. Average achievable secrecy rate R_{Sk} for each tier k of the network is given by (14).

5 Simulation Results and Performance Analysis

In this section, numerical results are shown to study and understand the impact of massive MIMO large antenna arrays and mmW channel characteristics on the secrecy outage probability, SSE, and SEE of the network. The simulation parameters are outlined in Table 2. Monte Carlo simulations are used to study the system performance.

We begin by studying the performance of the proposed network's secrecy outage probability in terms of density of eavesdropper nodes, as shown in Fig. 2. This plot has been obtained using (21) while taking varying small cell BS densities and directional antenna gains at the eavesdroppers nodes of the network. The secrecy outage probability of the network increases with increasing eavesdropper's density, thereby notifying that a large number of eavesdropping nodes in the network harm the network secrecy. However, there is another important observation to be noted that higher small cell BS density optimizes the secrecy of the network even though the eavesdropper's density might be high. These results allude to the fact that higher small cell density results in the increase in interference in the network. Thus, the uncertainty at eavesdropper nodes elevates leading to their SINR falling below the threshold. This results in improvement in secrecy outage probability of the network. The reader may notice that when the small cell density in the network is kept fixed,

Table 2 Simulation parameters

Parameter	Value	Parameter	Value
λ_1	$(500^2 \times \lambda)^{-1}$	$f_1 = f_2$	1 GHz
$B_1 = B_2$	10 MHz	α_1	3.5
α_2	4	P_1	46 dBm
P_2	30 dBm	λ_e	1×10^{-6}
f_3	28 GHz	B_3	100 MHz
P_3	30 dBm	α_L	2
α_N	4	N_N	2
N_L	3	M_r	10 dB
M_t	10 dB	m_r	−10 dB
m_t	0 dB	θ_r	90°
θ_t	30°	σ^2	−90 dBm
Noise figure	10 dB	$1/\beta$	141.4 m [30]
$\epsilon_1 = \epsilon_2 = \epsilon_3$	0.38 [28]	ρ_1	4 W
Δ_1	4.8	Δ_2	0
Δ_3	2.08×10^{-8}	Λ_1	1
Λ_2	9.5×10^{-8}	Λ_3	6.25×10^{-8}
ρ_2	13.6 W [39]	ρ_3	13.6 W [39]

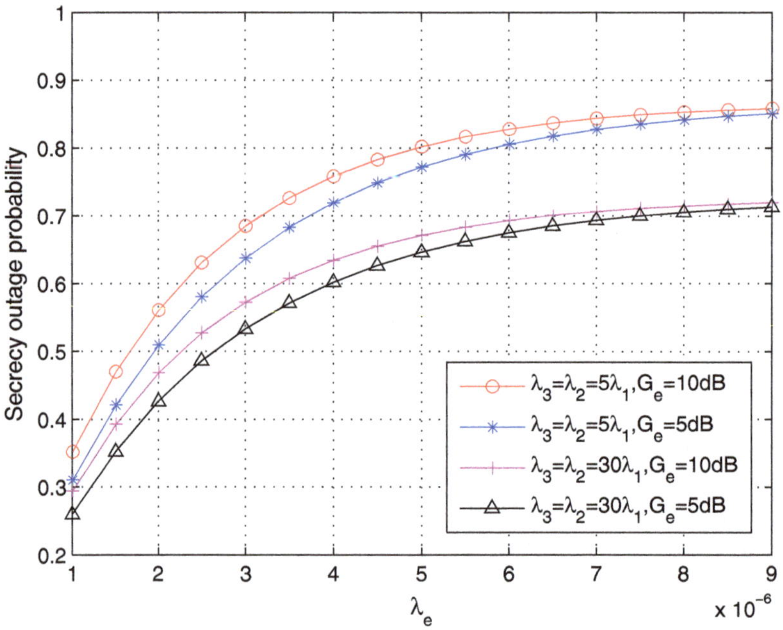

Fig. 2 Secrecy outage probability of the three-tier network as the function of λ_e for $\gamma_e = 40$ dB, $N = 5$

the secrecy rate of the transmissions in the network improves with small directional antenna gains at the eavesdroppers. We conclude that lower directional gains at the eavesdroppers and higher cell density in the network are the two major settings

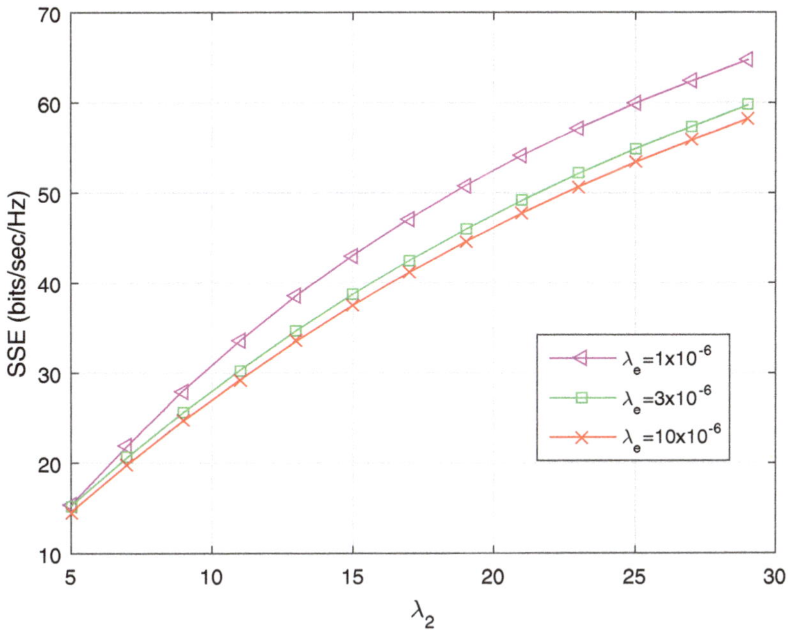

Fig. 3 SSE versus varying small cells BS density as multiple of MBS density for $\lambda_3 = \lambda_2$, $G_e =$ 15 dB, $N = 40$, $U = 5$

that reduce the probability of eavesdroppers having SINR above the predefined threshold.

Now, we study the SSE behavior based on varying small cell BS density of tier 2 and tier 3 for various eavesdropper's densities. Figure 3 plots the SSE of the network versus varying small cell densities. It can be seen that the SSE increases with the increase in small cell density and falls when the eavesdropper node's density increases. This is because, the average cell radius decreases when small cell density is increased. Therefore, more users are offloaded to small cells so the transmission from the BS to the intended legitimate user becomes better. When the cell sizes decrease, the distance from the intended receiver to the transmitting nodes decreases that causes stronger links and enhanced secrecy rate between them. If we consider the case of mmW small cells, the LoS probability function $p(R)$ is directly related to distance; thus, the LoS association probability increases with decrease in distance. In such scenario, low path loss LoS links between mmW BSs and legitimate users are more likely to be formed than NLoS links. Another interesting observation is that user's association with MBSs declines in the presented scenario though they are high-power nodes but have low BS density compared to the small cells. Interference in the network increases as transmitting node density is notably high; therefore, an increase in eavesdropper's density has little impact on SSE of the network.

Figure 4 illustrates the relationship between the SEE and the small cell density. This plot has been obtained using (23). As a preliminary observation, it may be

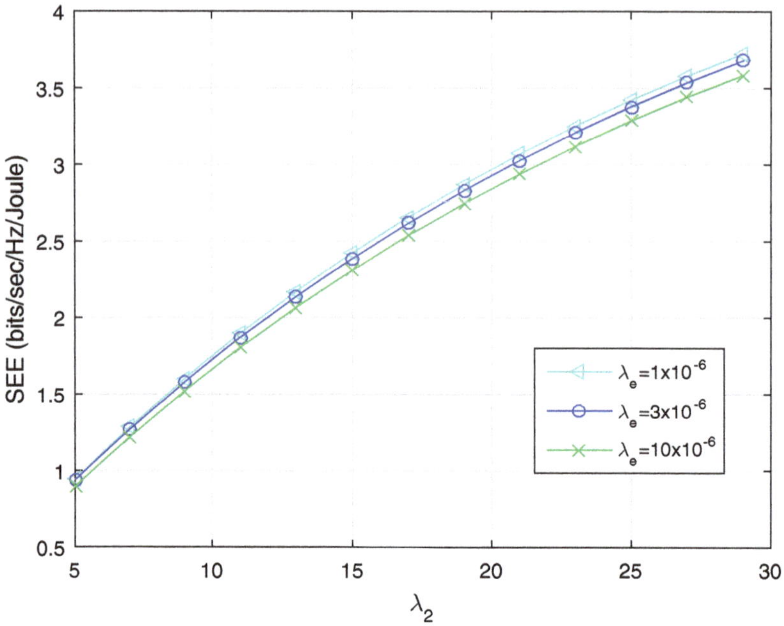

Fig. 4 SEE versus varying small cells BS density as multiple of MBS density for $\lambda_3 = \lambda_2$, $G_e = 15\,\text{dB}$, $N = 30$, $U = 5$

stated that the SEE is higher for greater values of λ_2 and λ_3, respectively . This is because of the increase in SSE over the identical power consumption, shown in Fig. 3. Moreover, on increasing small cell densities, λ_2 and λ_3, more users associate with low-power small cell BSs compared to high-power macro cell BSs. This shift in association leads to a power-efficient network; however, it comes at the cost of small cell BS deployment. Here again it may be noted that the interference in the three-tier network is higher with dense transmitting BS nodes; therefore, an increase in eavesdropper's density has little impact on SEE of the network.

Extending the previous discussion, we now vary the number of transmitting antennas on the massive MIMO-enabled MBS of the network and then observe the effects on SSE of the three-tier hybrid HetNet over different small cell densities. The results are shown in Fig. 5. It can be seen that the SSE of the network falls with the increasing number of antennas at the MBSs and shows significant increase with higher small cell density. This is because, when N increases, spectral efficiency of MBSs increases. As the MBSs are high power nodes, transmission to eavesdroppers in the network improves significantly that results in increases secrecy rate, R^e. Legitimate user to BS association gets biased toward MBSs because of their high array gains at increased number of antennas. Thus, fewer portion of users connect with small cells BSs. Hence, interference in the network reduces leading to better reception at eavesdroppers.

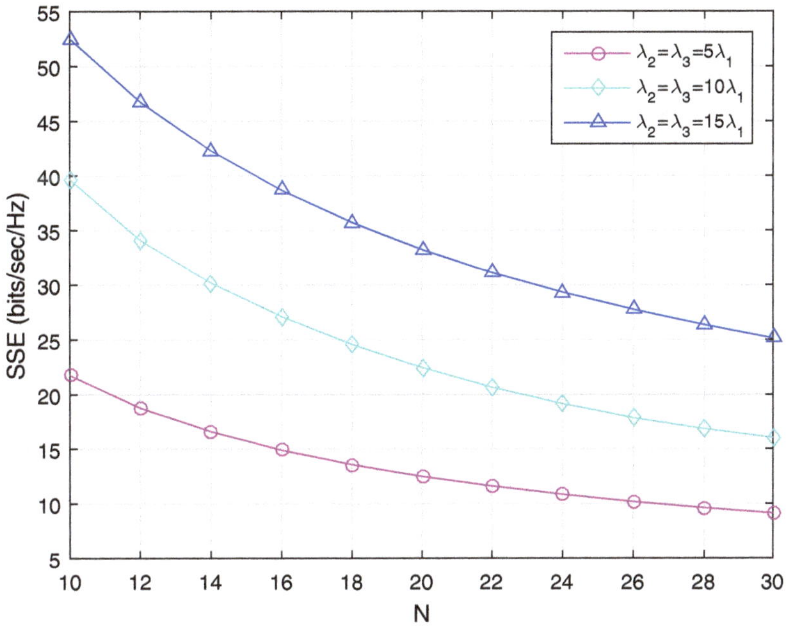

Fig. 5 SSE versus different number of antennas at macro BS for $G_e = 15\,\text{dB}$, $U = 5$

Figure 6 illustrates the variation in SEE with increasing number of antennas at massive MIMO-enabled MBSs. Plot is based on (23). It can be seen that SEE of the network drops with increasing N. It follows from the previous result that SSE drops over the identical power consumption. In addition, users are more likely to connect to macro cell BSs with increase in transmitting nodes at their BSs, and since macro cells are higher-power-consuming entities compared to small cells, therefore, overall network setting shifts to a power-inefficient network. As a result, SEE of the network decreases with the increase in average achievable rate at eavesdroppers. The reader may notice that the interference in the network elevates with small cell density; therefore, SEE improves with drop in eavesdroppers' secrecy rate R^e.

Early on, we mentioned that mmW and massive MIMO technologies are the candidates for being the enabling technologies for 5G networks due to their higher gains and superior bandwidth as for mmW technology. We conclude the Results section by highlighting these very important points that we cannot increase the directional beamforming gains at mmW nodes neither can we increase the number of antennas at massive MIMO-enabled nodes limitlessly to have optimum coverage in the proposed network, without entertaining drop in the overall secrecy performance of the network.

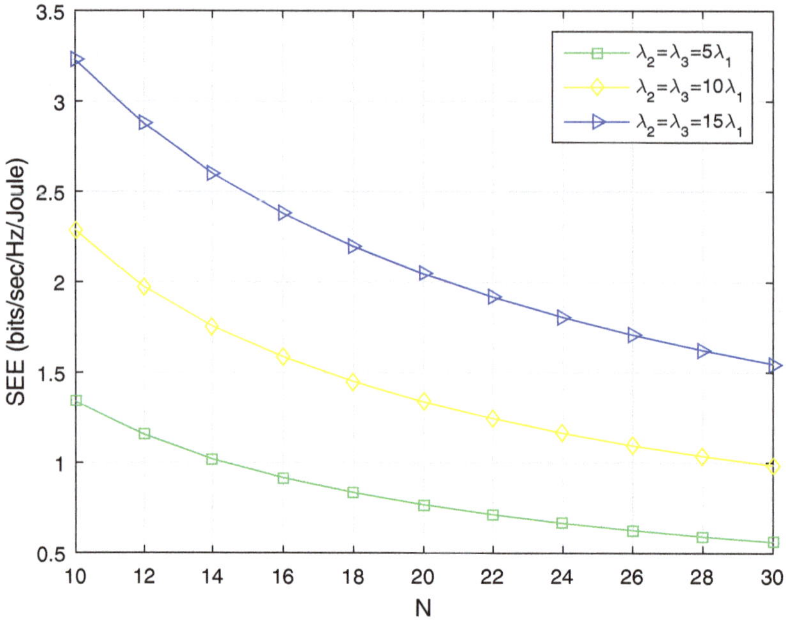

Fig. 6 SEE versus different number of antennas at macro BS for $G_e = 15\,\text{dB}$, $U = 5$

6 Conclusion

In this chapter, we characterized the secure communication in massive MIMO-enabled three-tier hybrid HetNet based on the unique features of massive MIMO and mmW communication by using PLS. Particularly, we evaluated the secrecy outage probability and achievable ergodic rate at each of the three network tiers and most malicious eavesdropper node. The expressions were then used to develop a tractable approach to determine network-wide SSE and SEE. It was observed that network-wide interference elevates on high BS density that in turn dominates the secrecy performance of the network. Moreover, the secrecy performance of the networks drops at higher beamforming gains at mmW cells; therefore, a tradeoff exists between optimal coverage and secrecy in the network. The relationship between the secrecy outage probability, SSE, and SEE of the network and number of antennas at MBS, BS density, antenna gains, and eavesdropper's density has also been studied through simulation results. Through these results, we conclude that the number of antennas at the massive MIMO-enabled macro cell BSs and beamforming directivity gains at mmW cells shall be carefully chosen for optimal secrecy performance of the network.

References

1. J.G. Andrews, S. Buzzi, W. Choi, S.V. Hanly, A. Lozano, A.C. Soong, J.C. Zhang, What will 5G be? IEEE J. Sel. Areas Commun. **32**(6), 1065–1082 (2014)
2. J. Zhang, X. Ge, Q. Li, M. Guizani, Y. Zhang, 5G millimeter-wave antenna array: design and challenges. IEEE Wirel. Commun. **24**(2), 106–112 (2017)
3. J. Ye, X. Ge, G. Mao, Y. Zhong, 5G ultra-dense networks with non-uniform distributed users. IEEE Trans. Veh. Technol. **67**(3), 2660–2670 (2018)
4. X. Ge, R. Zi, H. Wang, J. Zhang, M. Jo, Multi-user massive MIMO communication systems based on irregular antenna arrays. IEEE Trans. Wirel. Commun. **15**(8), 5287–5301 (2016)
5. R.W. Heath, T. Bai, R. Vaze, Analysis of blockage effects on urban cellular networks. IEEE Trans. Wirel. Commun. **13**(9), 5070–5083 (2014)
6. E. Turgut, M.C. Gursoy, Energy efficiency in relay-assisted mmW cellular networks, in *IEEE 84th Vehicular Technology Conference (VTC-Fall)*, Sept 2016, pp. 1–5
7. M. Ding, P. Wang, D. Lopez-Perez, G. Mao, Z. Lin, Performance impact of LoS and NLoS transmissions in dense cellular networks. IEEE Trans. Wirel. Commun. **15**(3), 2365–2380 (2016)
8. H. Wang, T. Zhang, X. Xia, Secure MISO wiretap channels with multi-antenna passive eavesdropper: artificial noise vs. artificial fast fading. IEEE Trans. Wirel. Commun. **14**(1), 94–106 (2015)
9. Y. Liang, H.V. Poor, S. Shamai, Secure communication over fading channels. IEEE Trans. Inf. Theory **54**(6), 2470–2492 (2008)
10. L. Wang, N. Yang, M. Elkashlan, P.L. Yeoh, J. Yuan, Physical layer security of maximal ratio combining in two-wave with diffuse power fading channels. IEEE Trans. Inf. Forensics Secur. **9**(2), 247–258 (2014)
11. H.M. Wang, M. Luo, Q. Yin, X.G. Xia, Hybrid cooperative beamforming and jamming for physical-layer security of two-way relay networks. IEEE Trans. Inf. Forensics Secur. **8**(12), 2007–2020 (2013)
12. T. Lv, H. Gao, S. Yang, Secrecy transmit beamforming for heterogeneous networks. IEEE J. Sel. Areas Commun. **33**(6), 1154–1170 (2015)
13. H. Wu, X. Tao, N. Li, J. Xu, Secrecy outage probability in multi-RAT heterogeneous networks. IEEE Commun. Lett. **20**(1), 53–56 (2016)
14. H. Wang, T. Zheng, J. Yuan, D. Towsley, M.H. Lee, Physical layer security in heterogeneous cellular networks. IEEE Trans. Commun. **64**(3), 1204–1219 (2016)
15. Y. Deng, L. Wang, K.K. Wong, A. Nallanathan, M. Elkashlan, S. Lambotharan, Safeguarding massive MIMO aided hetnets using physical layer security, in *Proceedings of Wireless Communications and Signal Processing*, 2015, pp. 1–5
16. J. Wang, J. Lee, F. Wang, T.Q.S. Quek, Jamming-aided secure communication in massive MIMO rician channels. IEEE Trans. Wirel. Commun. **14**(12), 6854–6868 (2015)
17. M. Xu, X. Tao, F. Yang, H. Wu, Enhancing secured coverage with COMP transmission in heterogeneous cellular networks. IEEE Commun. Lett. **20**(11), 2272–2275 (2016)
18. L. Wang, K.K. Wong, M. Elkashlan, A. Nallanathan, S. Lambotharan, Secrecy and energy efficiency in massive MIMO aided heterogeneous C-RAN: a new look at interference. IEEE J. Sel. Top. Sign. Proces. **10**(8), 1375–1389 (2016)
19. V. Sharma, D. N. K. Jayakody, I. You, R. Kumar, J. Li, Secure and efficient context-aware localization of drones in urban scenarios. IEEE Commun. Mag. **56**(4), 120–128 (2018)
20. V. Sharma et al., Secure and energy efficient handover in fog networks using blockchain-based DMM. IEEE Commun. Mag. **56**, 22–31 (2018)
21. F. Jameel et al., Secure communication for separated and integrated receiver architectures in SWIPT. IEEE Wirel. Commun. Netw. Conf. **2018**, 1–6 (2018)
22. L. Wang, M. Elkashlan, T.Q. Duong, R.W. Heath Jr., Secure communication in cellular networks: the benefits of millimeter wave mobile broadband, in *IEEE 15th International Workshop on Signal Processing and Advances in Wireless Communication (SPAWC)*, 2014, pp. 115–119

23. S. Vuppala, S. Biswas, T. Ratnarajah, An analysis on secure communication in millimeter/micro-wave hybrid networks. IEEE Trans. Commun. **64**(8), 3507–3519 (2016)
24. C. Wang, H.M. Wang, Physical layer security in millimeter wave cellular networks. IEEE Trans. Wirel. Commun. **15**(8), 5569–5585 (2016)
25. S. Gong, C. Xing, Z. Fei, S. Ma, Millimeter-wave secrecy beamforming designs for two-way amplify-and-forward MIMO relaying networks. IEEE Trans. Veh. Technol. Early Access Articles, **66**, 1–12 (2016)
26. A. Umer, S.A. Hassan, H. Pervaiz, Q. Ni, L. Musavian, Coverage and rate analysis for massive MIMO-enabled heterogeneous networks with millimeter wave small cells, in *IEEE 85th Vehicular Technology Conference (VTC Spring)*, 2017, Sydney, pp. 1–5
27. A. Umer, S.A. Hassan, H. Pervaiz, Q. Ni, L. Musavian, S.H. Ahmed, Secrecy outage analysis for massive MIMO-enabled multi-tier 5G hybrid hetnets, in *2018 IEEE International Conference on Communications Workshops*, 2018, Kansas City, pp. 1–6
28. Y. Hao, Q. Ni, H. Li, S. Hou, On the energy and spectral efficiency tradeoff in massive MIMO enabled hetnets with capacity-constrained Backhaul links. IEEE Trans. Commun. (2017, in press). https://doi.org/10.1109/TCOMM.2017.2730867
29. K. Hosseini, W. Yu, R.S. Adve, Large-scale mimo versus network mimo for multicell interference mitigation. IEEE J. Sel. Top. Sign. Proces. **8**(5), 930–941 (2014)
30. M. Omar, M. Anjum, S.A. Hassan, H. Pervaiz, Q. Ni, Performance analysis of hybrid 5G cellular networks exploiting mmW capabilities in suburban areas, in *IEEE Internation Conference on Communications*, Kuala Lumpur, 2016
31. E. Turgut, M.C. Gursoy, Coverage in heterogeneous downlink millimeter wave cellular networks. IEEE Trans. Commun. **65**, 4463–4477 (2017)
32. C. Yang, J. Li, Q. Ni, A. Anpalagan, M. Guizani, Interference-aware energy efficiency maximization in 5G ultra-dense networks. IEEE Trans. Commun. **65**(2), 728–739 (2017)
33. A.D. Wyner, The wire-tap channel. Bell Labs Tech. J. **54**(8), 1355–1387 (1975)
34. X. Zhang, X. Zhou, M.R. McKay, Enhancing secrecy with multi-antenna transmission in wireless ad hoc networks. IEEE Trans. Inf. Forensics Secur. **8**(11), 1802–1814 (2013)
35. C. Ma, J. Liu, X. Tian, H. Yu, Y. Cui, X. Wang, Interference exploitation in D2D-enabled cellular networks: a secrecy perspective. IEEE Trans. Commun. **63**(1), 229–242 (2015)
36. P.C. Pinto, J. Barros, M.Z. Win, Secure communication in stochastic wireless networks–part I: connectivity. IEEE Trans. Inf. Forensics Secur. **7**(1), 125–138 (2012)
37. T. Bai, R.W. Heath, Coverage and rate analysis for millimeter-wave cellular networks. IEEE Trans. Wirel. Commun. **14**(2), 1100–1114 (2015)
38. H. Pervaiz, L. Musavian, Q. Ni, Z. Ding, Energy and spectrum efficient transmission techniques under QoS constraints toward green heterogeneous networks. IEEE Access **3**, 1655–1671 (2015)
39. G. Auer, V. Giannini, C. Desset, I. Godor, P. Skillermark, M. Olsson et al., How much energy is needed to run a wireless network? IEEE Wirel. Commun. **18**(5), 40–49 (2011)

Anum Umer received the B.E. degree in Electrical(Telecommunication) Engineering from the National University of Science and Technology (NUST), Pakistan in 2015 and the M.S. degree in Electrical Engineering from NUST, Pakistan in 2017. Her broader area of research includes wireless communication, massive MIMO and millimeter wave communication. She is currently a Researcher with the System Analysis and Verification Lab, School of Electrical Engineering and Computer Science, NUST.

Syed Ali Hassan received the B.E. degree (Hons.) in electrical engineering from the National University of Sciences and Technology (NUST), Pakistan, in 2004, the M.S. degree in elec- trical engineering from the University of Stuttgart, Germany, in 2007, and the M.S. degree in mathe- matics and the Ph.D. degree in electrical engineering from the Georgia Institute of Technology, Atlanta, GA, USA, in 2011. His broader area of research is signal processing for communications. He is currently an Assistant Professor with the NUST School of Electrical Engineering and Computer Science, where he is also the Director of Information Processing and Transmission Research Group, which focuses on various aspects of theoretical communications. He was a visiting professor at Georgia Tech in Fall 2017 and also held industry position at Cisco Systems Inc, CA, USA.

Physical Layer Security of Energy Harvesting Machine-to-Machine Communication System

Furqan Jameel, Muhammad Awais Javed, and Dushantha Nalin K. Jayakody

1 Introduction to Machine-to-Machine Communications

The dawn of Internet of Things (IoT) has given birth to efficient methods of communication between devices. In fact, Ericsson predicts that there will be about 15 billion devices in 2021, and most of them will be low-power devices with short-range communication [1]. To achieve this landmark, one of the key enabling technologies would be machine-to-machine (M2M) communications. M2M communication is defined as the communication between two or more devices in which direct human intervention is not necessarily needed [2]. These networks are generally installed for industry production automation, smart housing, and environmental monitoring. The name machine-type communications (MTC) or M2M communications has been given by 3rd generation partnership project (3GPP). One of the critical issues faced by M2M communication is safely connecting an enormous number of machine-type communication devices (MTCDs) to the web.

The existing mobile cellular systems are expected to evolve in high-density access points (APs) and distributed antenna system (DAS). Thus, the mechanism to maintain communication between machines in the dual designs like small cells and DAS is still an open research issue. The user devices, entries, and base station (BS) can be used as a relay in-between MTCDs and M2M servers. Routing setups are

F. Jameel (✉)
Faculty of Information Technology, University of Jyväskylä, Jyväskylä, Finland

M. A. Javed
Department of Electrical Engineering, COMSATS University, Islamabad, Pakistan

D. N. K. Jayakody
School of Computer Science and Robotic, National Research Tomsk Polytechnic University, Tomsk, The Tomsk Area, Russia

School of Postgraduate Studies, Sri Lanka Technological University, Padukka, Sri Lanka

D. N. K. Jayakody et al. (eds.), *5G Enabled Secure Wireless Networks*,
https://doi.org/10.1007/978-3-030-03508-2_5

also critical issues and need to be resolved along with the appropriate allocation of cellular resources by formulating reliable and secure communication mechanisms.

1.1 Applications of M2M Communications

As shown in Fig. 1a, M2M communication architecture consists of three domains, namely, M2M domain, network domain, and application domain. Among these

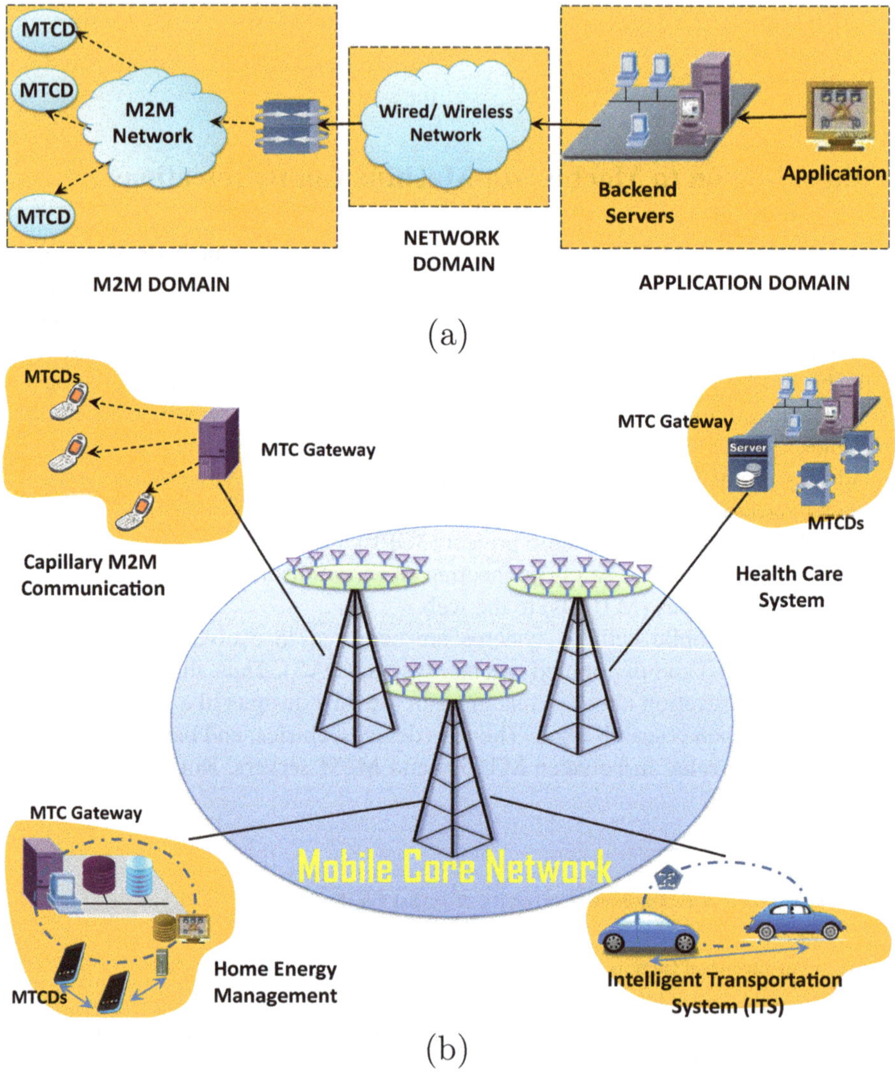

Fig. 1 M2M communications systems (**a**) M2M architecture showing M2M, network and application domains (**b**) M2M applications that include energy management, healthcare, and intelligent transportation system

domains, MTCDs mainly exist in the M2M domain and connect to the rest of the network through MTC gateways. The MTC gateways provide certain computational capabilities and process the data from other MTCDs using short-range transmission [3]. These gateways also prevent BSs in network domain from getting overwhelmed and act as relaying devices between MTCDs and BSs. Some major applications of M2M communications have been shown in Fig. 1b. In the field of healthcare, M2M communications have several applications which include wireless body area networks (WBANs) by interconnecting multiple short-range communication devices [4]. Particularly, an M2M-aided gateway node gathers the complete sensor statistics from the network and directs that data to a remote online server. Here, the task of managing the health-associated software requests is accomplished. The server is also connected to the web with the help of technologies like wireless local area networks (WLANs) and wireless metropolitan area networks (WMANs). The characteristics of M2M communications allow the distant monitoring of vigorous signs. The information about the critical condition of a patient and some medical treatment, done remotely, can also be performed [5]. It can also facilitate reliable data exchange between patient and hospital monitoring units. The authors of [6] suggested the applicability of M2M communications in smart grid systems. They showed that MTCDs can transfer information to the end user by using low- powered communication equipment. Automotive vehicular networks can also greatly benefit from M2M communications. Specifically, vehicle telematics and remote traffic management can be performed using M2M communications [7]. Other applications include automation system in buildings, surveillance and security systems, multimedia distribution networks, and fleet management systems [8]. It is worth highlighting that these applications have different quality of service (QoS) requirements in terms of their immediacy and transmission frequency. These requirements have been summarized in Table 1.

Table 1 Classification of QoS requirements of M2M applications

M2M application	Immediacy/Priority	Transmission cycle	Packet length
Smart grid	Medium, due to large volume of packets to collection centers	Periodic	Long
Railway systems	High, due to transmission of real-time alert messages	Periodic	Short
Mobile streaming	Medium, due to continuous transmission of video messages	On-demand	Long
E-health services	High, due to transmission of sensitive and critical data	Periodic	Short
Surveillance/ Monitoring	High, due to transmission of continuous data to monitoring centers	On-demand	Short
Emergency services	High, due to bursty traffic demands	Random	Short/Long
Navigation system	Medium, due to exchange of real-time data	Random	Short

1.2 Design and Performance Analysis of M2M Communications

High-speed transfer of data has become an indispensable part of day-to-day life. In fact, the growing demand for data exchange has highlighted some key limitations of existing wireless networks. The present cellular infrastructure is primarily designed to support mobile devices having random transmission behaviors and mobility patterns. In contrast, M2M communications follow a specific duty cycle for information transfer, and it is delay tolerant with the ability to transfer small data packets between stationary terminals [9]. Thus, it can be deduced that connecting a large number of MTCDs with cellular networks can saturate uplink data transmission [3]. In this context, noncellular connections can serve as an important supplement to cellular networks. Noncellular networks can be classified into the following two types:

1. Widely used wireless local/personal area networks
2. Emerging low power wide area (LPWA) networks

The representative technologies for local/personal area networks are WLANs, smart utility networks (SUNs), Zigbee, Bluetooth, and ultra-wideband (UWB) networks. The analysis on the aforementioned networks has been carried out in the literature. For instance, a comparative analysis was presented by GreenPeak in [10] and by Chen et al. in [3], while experimental solutions were suggested by Keysight. Among these techniques, UWB is different from the other wireless technologies, which generates little interference for narrow-band signals and uses a large bandwidth. UWB is suitable for demanding indoor applications, such as video cameras and surveillance. For most applications of M2M communications, transmissions over narrow-band channels are sufficient to meet communication requirements. Also, devices may need to operate with a low power consumption and work for a considerably long period with batteries.

1.3 M2M Security Challenges and State-of-the-Art Solutions

The abovementioned applications of M2M communications can only be feasible if the issue of security is properly addressed. Although security is a critical aspect of M2M communication, different aspects and characteristics of MTCDs may create challenges to design an efficient and secure communication mechanism. Several key contributions have been made in the security domain of IoTs [11], device-to-device (D2D) communications [12], and wireless sensor networks (WSNs) [13]; still the M2M paradigm requires dynamic and energy efficient security solutions for its long-term implementation.

In this regard, the authors of [14] present recent advances on architecture and standardization of the M2M system. The main focus of the said work was on security issues due to the heterogeneous nature of the devices. This, in turn, raises some serious problems as devices of different vendors having different technologies have to coexist in the same environment. Mainly, the following are concerning issues:

- How to identify the devices in an ultra-dense network setup?
- How to avoid information leakage to unauthorized devices/users?

In order to deal with the first problem, the same authors propose a unique identifying mechanism. The said mechanism allowed devices to identify each other despite difference in their technologies. For the second problem, the authors analyzed different key establishment techniques and define a basic process for secure transmission on the data.

In another work [15], Lu et al. presented two mechanisms, namely, bandwidth efficient cooperative authentication (BECAN) and early detection node. The main objective of both schemes was to filter the unauthentic and false data. In the BECAN scheme, the false data can be detected by filtering through an en route node. The filtering is performed based on two metrics, namely, filtering ratio and en route filtering probability. For the case of the early detection mechanism, the malicious node is detected by coupling with its neighboring nodes. The authors have shown that the en route filtering probability increased with an increase in the number of en route nodes in the network.

Kitagami et al. in [16], proposed a long polling communication method to satisfy the requirements of security and immediacy. A load-balancing server was introduced by the authors that increase the duration of communication sessions among devices. The performance of the said scheme was evaluated against two performance metrics and load index. The simulation results presented by the authors validated that the load balancing mechanism functions efficiently.

In addition to above-stated works, some other techniques have also been proposed in the literature to offer further insights into the security of M2M networks. The authors of [17] unveiled that MTCDs mainly operate in sleeping (energy saving) or operating (data exchange) modes. However, the unattended sleeping MTCDs may be most vulnerable to cyberattacks. Moreover, during a wake up cycle, these compromised nodes may inject false information into the network. Due to this reason, it is essential to secure and monitor the unusual events in the network. These unusual events may include irregularities in sleep-wake cycles of devices, rapid change in the location of MTCDs, and frequent occurrence of hardware-related issues [18]. Other privacy and security issues include protocol attacks, configuration attacks, identity attacks, and physical attacks [19].

2 Energy Harvesting

Energy harvesting (also called power harvesting) is a technique through which energy is scavenged from external sources (e.g., solar power, thermal energy, wind energy, salinity gradients, and kinetic energy) [20]. Energy harvester gives a

Table 2 Energy harvesting estimates by Texas Instruments [22]

Energy source	Harvested power
Vibration/motion	
Human	4 μ W/cm^2
Industry	100 μ W/cm^2
Temperature difference	
Human	25 μ W/cm^2
Industry	1-10 μ W/cm^2
Light	
Indoor	10 μ W/cm^2
Outdoor	10 mW/cm^2
RF	
GSM	0.1 μ W/cm^2
WiFi	0.001 mW/cm^2

little supply of power which is consumed by low-energy electronics. The energy providing source for the energy harvesters is mostly available in a large-scale environment and is free of cost. For example, due to the radio, cellular, and television broadcasting, electromagnetic energy is always available and is free to use for harvesting. The concept of energy harvesting has been around for many years as it is persistently evolving. Energy harvesting has several advantages as it is mostly infinite and available in bulk, and the process is generally environment-friendly with minimum maintenance cost [21]. Texas Instruments studied statistics of various sources to determine the amount of harvested energy, as shown in Table 2.

2.1 Energy Harvesting Sources

The assorted sources of energy harvesting are photovoltaic cells, solar energy, wind turbine, electromagnetic waves, and mechanical vibration devices like piezoelectric and electromagnetic devices. At present, the main sources of ambient energy which are suitable for use in wireless devices are RF energy, solar, wind energy, thermal, and vibration energy [23, 24]. Solar power is the most mature form among other harvesting sources, but it only generates energy in presence of sunlight or artificial light which is a drawback of solar power [25].

There are different ways of harvesting energy from the energy sources. For instance, in an RF energy harvesting system, radio waves from a nearby cellular tower are used to exploit the dual nature of RF waves. The voltage level stored in the diode is then boosted to a higher level and stored in a supercapacitor. Thermal energy uses temperature gradient to produce electricity from both the human body and surrounding environment [26]. Thermal energy harvesting system is integrated with tiny devices. Thermoelectric generator (TEG) is used to convert human body heat into DC voltage and is stored in a supercapacitor. The resultant harvested DC voltage by these sources is then boosted to a higher voltage level and is used by the sensors. As a result, sensors may operate without changing batteries which saves time and money and increases the reliability of a sensor network.

Rechargeable batteries are modeled as energy buffer in the energy harvesting sensor network, where scavenged energy is stored with respect to charging characteristics of the battery. In energy harvesting wireless networks, energy management policy is taken into consideration in energy refilling process. Since the energy supply is random and uncertain, the design consideration of harvested energy system is different as compared to the non-rechargeable battery system. Moreover, energy management policy can be taken into consideration for energy refilling process and for increasing the lifetime of wireless devices [27]. The definition of the lifetime of devices varies and mostly depends on the application and network topology. Some of the major definitions are appended below:

1. Time period during which certain numbers of nodes have consumed batteries [28].
2. Lifetime of the device/node which has maximum energy consumption rate [29].
3. Time when first data collection failure occur [30].
4. Time duration when first node of network is unavailable [31].

Although energy harvesting is a vast topic, however, our focus here is toward RF energy harvesting as it is one of the emerging paradigms for future 5G communications [32]. Moreover, RF signals can be used to transfer both information and energy to a remote user, thus having dual use in modern communication systems.

2.2 RF Energy Harvesting

Due to rapid advancements in communication technologies, RF signal is broadcasted from several sources including radio transmitters including BSs, handheld phones, cellular phones, television broadcasting terminals, etc. The capability of harvesting RF energy from ambient or dedicated sources can revolutionize the usage of existing mobile device [20]. An efficient scheme can theoretically remove the dependency of battery charging or its restoration. In either case, devices become free of wires, connectors, and battery entry board, thus providing much-needed portability in the process of charging. Low-powered devices can be charged from the nearest range of dedicated power transmitter [32]. Moreover, this energy can be used to charge many devices simultaneously which may include global positing system (GPS), wearable medical sensor, and end-user electronics like e-book readers and headsets. In a highly populated urban area, electromagnetic waves are easily available over a wide spectrum of frequencies and at different potential levels. As the power consumption of electronics components is continuously increasing, the demand of applications like thread-free charging with the help of RF harvester will grow in the future.

Recent Developments in Wireless Power Transfer

In recent years, wireless power transfer for powering mobile terminals began to attract increasing research interest [33, 34]. Wireless power transfer has been performed using free-space beamforming and antennas with large apertures in order to overcome propagation loss for large power transfer. Huang et al. in [33] proposed an overlaying uplink cellular network architecture for RF charging stations. In [34], a harvest-then-transmit protocol was introduced, and different beamforming techniques were employed to improve power transfer efficiency for mobile applications. It is until recently that the dual use of RF signals for delivering energy as well as for transporting information has been advocated [35]. Particularly, simultaneous wireless information and power transfer (SWIPT) technology have been proposed as a potential candidate for transmission of information and energy over the same frequency band. It provides the advantage of delivering controllable and efficient on-demand wireless information and energy at a low-cost option for sustainable operations of wireless systems. Due to this reason, SWIPT has several applications in 5G networks including cooperative beamforming, ad hoc communications, D2D communications, etc. [36], as illustrated in Fig. 2.

Recent research has shown that optimizing wireless information and energy transfer simultaneously brings trade-off with respect to the design aspect of wireless

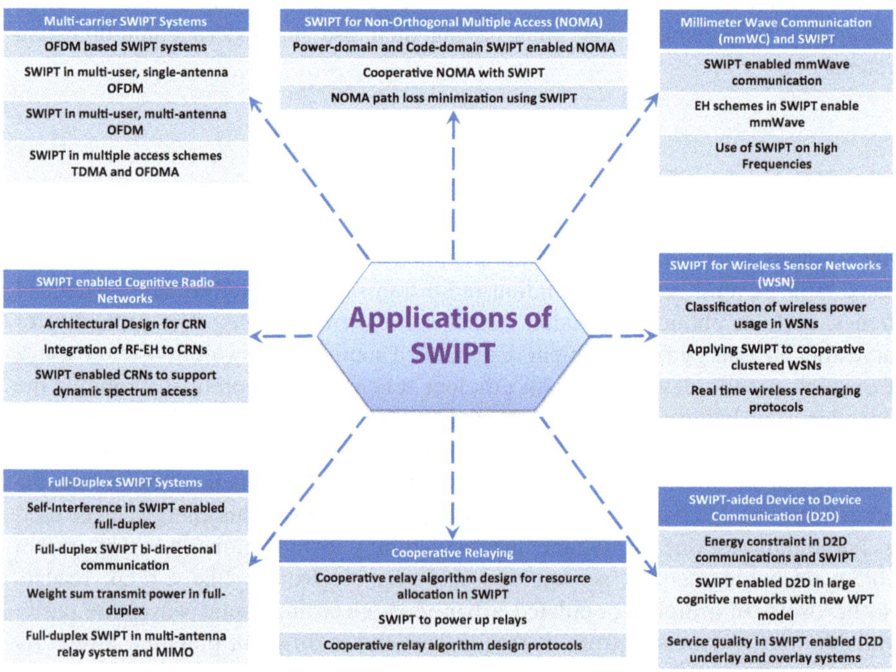

Fig. 2 Common applications of SWIPT in wireless communications

systems [37]. Thus, the majority of research efforts nowadays are directed toward receiver operation designs. Various operation policies have also been proposed for multiuser downlink systems with single-input single-output (SISO) channels [38] and multiple-input single-output (MISO) channels [39]. The work of [38] investigates an orthogonal frequency division multiple access (OFDMA) system with SWIPT from an energy efficiency perspective. Beamforming design problems for a two-way relay system were studied by the authors of [40]. Two single-antenna source nodes exchange information through multiple relay nodes and harvest RF energy from the transmission of these relays. In a similar work [41], Michalopoulos et al. investigate the relay selection techniques under Rayleigh-faded network conditions with a separated information receiver and energy harvester.

3 Principles of Physical Layer Security

In almost all forms of wireless communication, the issue of security and confidentiality is dealt at the upper layers of the protocol stack. Such security measures assume that the physical layer is error free. The mathematical models generate keys for such security mechanisms and are called computational security. Even though computational security has proven itself to be a very fruitful mechanism, the variety of emerging network architectures (ad hoc networks) requires additional security measures [32]. The information security techniques using shared key assume that the channel is perfect which is not the case in practice, and if either the key or information is changed, security will be lost. The idea of securing information by harnessing the randomness of the channel characteristics has turned into a research area called physical layer security (PLS) [42]. PLS utilizes the randomness of the wireless channel to ensure reliable and secure communication. While traditional security techniques are limited by the processing power of the wireless nodes, PLS can achieve communication secrecy without requiring computationally expensive cryptographic operations [43]. Using PLS, message confidentiality and authentication can be achieved in the presence of malicious eavesdroppers. A list of recent research trends in PLS is given in Table 3.

3.1 Categorization of Eavesdroppers

Eavesdroppers can be categorized into the following two major categories:

Based on Cooperation

In case of multiple adversaries, eavesdroppers may work independently or cooperatively.

- Non-colluding eavesdroppers are mutually independent and do not share received information to cooperatively decode the confidential message.

Table 3 Recent research trends in PLS

Security issue	Reference	Network type	Solution
Key agreement	[44]	Mobile networks	Deep fade detection for randomness extraction; Light-weight information reconciliation
Authentication	[45, 46]	Wireless network	Fingerprinting
	[47]	Wireless body area networks	Wireless channel exploitation
	[48]	Mobile network	Time varying carrier frequency offset
	[49]	Cognitive radio networks	Authentic tag generation by one way hash chain
Secrecy capacity enhancement	[50, 51]	Cooperative wireless network	Secrecy optimization
	[52]	Cognitive radio networks	Cooperative jamming
	[51]	Cellular networks	Stochastic geometry and random matrix theory

- For colluding eavesdroppers, multiple adversaries try to intercept the communication and mutually share the information such as received SNR, to decode the message.

Based on Activity

An eavesdropper can try to intercept or jam the communication between legitimated users.

- Radio eavesdroppers, also called passive eavesdroppers, are capable to detect and intercept the main transmission without bringing any changes in the network. Also, they cannot make any modification in the message that is received at intended receiver. Due to this reason, these types of attacks are difficult to detect.
- Active attacks can intercept and monitor the transmission and have the capability to bring modifications in the main channel. The major aim of this type of attack is to degrade the received SNR at the intended receiver which causes more decoding errors.

3.2 *Comparative Analysis of Secure Energy Harvesting Protocols*

It has now been well established that wireless signals can be used to carry both information and energy. However, the band allocated for information transmission cannot be used for power transfer [53]. Resultantly, the reduction in bandwidth

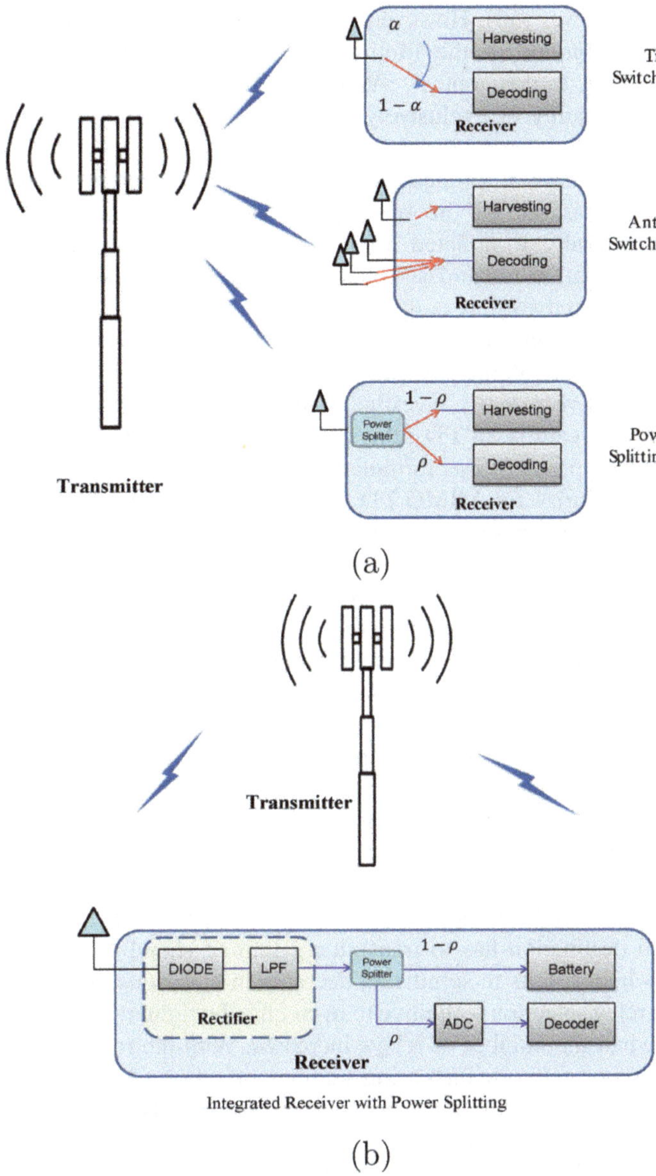

Fig. 3 SWIPT receiver architectures (**a**) Separated architecture (**b**) Integrated architecture

undermined the information transmission capability of devices. The separated and integrated receiver architectures of SWIPT [54] and techniques like time switching (TS) and power splitting (PS) were introduced for information and energy transfer, as shown in Fig. 3. Information transfer over relays has also been actively researched area of communication as it holds the promise to increase coverage area without increasing transmit power [55]. However, the performance of these relay-assisted networks is mostly limited to the lifetime of relays especially when constant power is not available. In this context, SWIPT has been applied to wireless relays whose operation is mainly or exclusively dependent on the amount of harvested energy [56].

Due to broadcast nature of wireless networks, secure transmission of data over the wireless channel is one of the fundamental issues. The design of SWIPT systems can be naturally formulated as a dual-objective optimization problem, one for secure information transmission, the other for efficient power transfer. In general, these two objectives may conflict with each other. If a power receiver is a potential eavesdropper, then the effort in improving the power transfer efficiency via increasing the power of information signal may result in a loss of secrecy rate. A few studies have recently investigated the PLS for energy harvesting full duplex (FD) relays. Security in FD relays with SWIPT systems was studied in [57]. An optimization problem was formulated for minimizing power consumption and ensuring link security for MIMO FD relays. In addition to this, the authors in [58] proposed strategy based on energy accumulation and then jamming the eavesdropper reception. Secrecy performance was evaluated for two modes of system operation: (1) opportunistic energy harvesting and (2) dedicated energy harvesting. Wu et al. in [59] maximize the secrecy performance of FD energy harvesting relays by solving the non-convex problem using Lagrange dual method.

The authors of [60] suggested two-stage cooperative jamming procedures to improve the secrecy rate of a wireless communication system. The key benefits of the scheme were the following: (1) global information is not required for the helpers, which makes the scheme more practical; and (2) secure performance is better even when the malicious user's channel quality is better than the main communication link. The same authors have shown that their scheme works well even without the eavesdropper's channel state information (CSI). In [61], the authors have studied the transmission strategy in a distributed relay networks for PLS and proposed a scheme in which destination has to transmit a jamming signal to complicate the malicious user, which results in significant decrease in malicious user SINR. They have proposed a relay selection technique to overcome the tighter secrecy constraint in the first slot when the number of relays increases. With the relay selection, the secrecy rate improves with an increasing number of relays. A power allocation method among the first/second slot and jamming/data signals has been proposed to further improve the secrecy rate.

The authors of [62] proposed relay selection methods to overcome the tighter secrecy constraint under known eavesdropper CSI and unknown eavesdropper CSI when the relaying nodes are increased. Using the proposed relay selection scheme, the secrecy rate is improved even with a small increase in the number of relaying devices. The authors of [63] proposed a transmission strategy in which destination transmits a jamming signal to confuse the eavesdropper thus improving the secrecy rate. In this scheme, the system gets benefit in the form of increment in total transmission power, and secrecy rate improves when the number of relays is increased. Proposed optimal power allocation among the first/second slot and jamming/data signals helps to enhance the secrecy rate. They proposed a distributed limited feedback scheme to reduce the feedback payload. Results have shown that the system can avoid a secrecy rate ceiling until feedback bits increase beyond a pre-specified rate for a fixed number of relays.

In [64], the authors analyzed a nulling noise scheme in which each helper uses its own link to the destination and independently transmits noise which leaks no interference to the destination. They proposed an optimal jamming noise structure and showed that nulling noise has secrecy rate performance very close to optimally designed noise with considerably small CSI requirements. Closed form of outage probability for nulling scheme is provided based on the statistics of malicious user CSI. They used a single antenna for the transmitter, the legitimate receiver, and malicious user, while helper has multiple antennas. To perform nulling at the destination, the helper must have more antennas than the destination. In [65], the authors proposed a cooperative jamming scheme for multiple helpers transmitting nulling noise to maximize the secrecy rate depending power under outage probability constraint. They formulated and solved outage-constrained secrecy rate maximization problem. It was considered that only transmitter has information about its channel to the legitimate receiver, while CSI to the channel of the eavesdropper is unknown. The proposed cooperative jamming scheme outperformed the existing benchmarking scheme.

In [66], the authors analyzed PLS of a bi-hop communication model constituting several AF relays and an eavesdropper. Their aim was to maximize the secrecy rate while efficiently allocating resources under certain constraints. They considered perfect knowledge of CSI such that base station (BS) had the knowledge of all the channels between BS and the eavesdropper. However, the CSI from BS to the users and from the BS to eavesdropper was unknown at the relay due to a large distance. All the eavesdroppers receive information message through relaying nodes. Essentially, the subcarrier allocation was used to allocate resources to individual users. The authors proposed a relay assignment and power allocation strategy over different subcarriers at the transmitting nodes. A suboptimal solution was derived and was shown to achieve significant secrecy performance gains as compared to conventional solutions. Results have also shown that the proposed dual-hop strategy provides performance improvement for different values of network parameters.

If artificial interference is added in a controlled manner, it can make a noteworthy difference between the way signal is interpreted at legitimate receiver and at the eavesdropper. Jamming can be used to enhance the PLS by performing antenna selection along with artificial interference [67]. Here, the authors consider a network consisting of a source, a destination, an eavesdropper, and a single relay. All the nodes have multiple antennas, and the artificial interference is added. The secrecy performance of the system is evaluated again and again by adding more interference to the point where the secrecy rate is considerably improved. It has been shown through the simulations that the information leakage to the eavesdropper reduces as the numbers of antennas are increased.

Optimum power allocation between data and interference is a critical aspect in PLS; therefore, the authors of [68] provided a robust artificial interference generation technique under channel uncertainty. Generally, they focused toward downlink multiuser scenarios and derived closed-form expressions for optimum power. The eavesdropper acts passively and the users, as well as eavesdroppers, have perfect knowledge of CSI of all the links. The signal is the first precoded using any one of the three precoding techniques including channel inversion (CI), zero-forcing (ZF), and regularized channel inversion (RCSI). Here, the authors found that the secrecy performance of RCI is better than the others. It was also shown that the secrecy rate decreases with an increase in the ratio of the large-scale path loss of main and wiretap channels. The authors also note that the secrecy rate becomes independent of path loss exponent at higher values of transmitting power.

When the CSI of the eavesdropper is not known at the receiver, the authors of [69] improved the secrecy performance of the system by transmitting an interference signal from the receiver. Similarly, the secrecy performance of a MIMO precoding system under optimum power allocation was improved in [70]. The authors derive the closed-form expression of the power allocation factor for maximizing the secrecy rate. They concluded that the tightness of bounds depends upon the number of transmitting antennas. The authors showed that the artificial interference can be added to impair the eavesdropper channel, and the scheme was called "mask beamforming." The proposed scheme divides the total power between noise and signal in order to ensure that a positive secrecy rate is achieved even if the eavesdropper's noise variance approaches 0. The simulation results have shown that as the transmitter antennas increase, the secrecy rate also increases proportionally.

4 Secrecy Performance of Energy Harvesting M2M Networks

We now provide secrecy performance analysis of an energy harvesting M2M communication system. In particular, we derive the closed-form expression of intercept probability of destination-assisted jamming technique under Rayleigh fading. We consider that imperfect channel knowledge is available at the transmitter; thus previously derived results for destination-assisted jamming are the special case of our system setup. For comparative analysis, we consider direct transmission as a benchmark method and also assume imperfect channel knowledge for this case.

4.1 System Model

This work considers a SWIPT-based transmission model, wherein a source S communicates with destination D in the presence of a single eavesdropper E. All the channels are assumed to be independent Rayleigh distributed. Fading remains constant for a single block of time and randomly changes from one block to another. This scenario can take place when eavesdroppers act as legitimate nodes for one transmission and wiretapping node for another. Both S and E are equipped with single antenna, whereas D is equipped with two isolated antennas [71].

Assumptions

This section explicitly states the definitions and assumptions used to facilitate the analysis in this chapter.

- We assume the case of reciprocal channels where the channel impulse response of uplink and downlink is alike. Furthermore, a feedback channel exists between transmitter and receiver and transmitter and eavesdropper which allows both parties to share the CSI.
- We consider the case of unicast communication where a transmitter sends a message to only one legitimate receiver. In this context, the receiver acts as a passive entity and does not send information in uplink. We also assume a block fading model in which channel remains unchanged during one block and changes randomly from one block to another.
- We also assume the case of passive eavesdropping in which eavesdropper cannot alter the information at legitimate receiver.
- We do not consider limited processing capability of the eavesdropper.

Dedicated Jamming

Let us first consider the benchmark scheme, as shown in Fig. 4, where D does not operate in FD mode, and, therefore, it does not transmit an interfering signal to E using its second antenna. We consider that both information decoding (ID) and energy harvesting receivers are co-located in D which uses time-switching (TS) protocol to decode information and harvest energy from RF signal. When S transmits the signal s to D, the transmission is also eavesdropped by E. At the same time, a helping interfering device I also transmit the interference signal to E. It is assumed that the interference signal can be canceled at D due to previously shared information regarding the jamming signal. The signal received at E is given as

$$y_E = \sqrt{\frac{P}{d_{SE}^{\chi}}}\hat{h}_{SE}s + \sqrt{\frac{P}{(1-\alpha)d_{IE}^{\chi}}}\hat{h}_{IE}s_I + n_E, \tag{1}$$

where d_{SE} is the distance between S and E, χ is the path loss exponent, $\hat{h}_{SE}$ is the estimated channel gain between S and E, and n_E denotes the additive white Gaussian noise (AWGN) with zero mean and N_0 variance which is normally

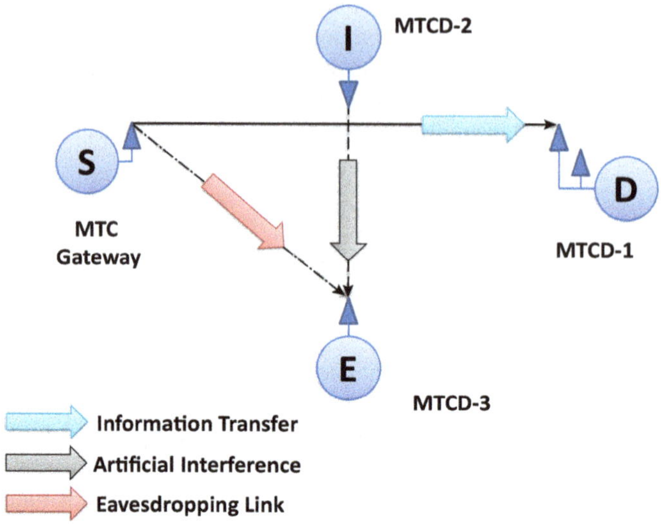

Fig. 4 System model for dedicated jamming

distributed as $\mathcal{N}(0, N_0)$. Also, $\hat{h}_{IE}$ is the estimated channel gain between I and E, d_{IE} is the distance between I and E, and s_I is the interfering signal at E. In practice, it is not possible to estimate the channel perfectly due to random nature of wireless channel and hardware impairments. Therefore, to adopt a more practical approach, we use the following channel error model for our analysis:

$$\hat{h} = \sqrt{1-\delta^2}h + \delta v, \tag{2}$$

where h is the actual channel gain and $0 \leq \delta \leq 1$ is the percentage accuracy of channel estimation. It is obvious that for $\delta = 0$, the estimation is perfect without any errors. In contrast, when $\delta = 1$, then channel estimation is completely inaccurate. Additionally, v is a complex Gaussian random variable with zero mean and unit variance. Replacing (2) in (1) yields

$$y_E = \sqrt{\frac{P(1-\delta_{SE}^2)}{d_{SE}^{\chi}}} h_{SE}s + \sqrt{\frac{P}{d_{SE}^{\chi}}}\delta_{SE}sv + \sqrt{\frac{P(1-\delta_{IE}^2)}{(1-\alpha)d_{IE}^{\chi}}} \times h_{IE}s_I + \sqrt{\frac{P}{(1-\alpha)d_{IE}^{\chi}}}\delta_{IE}s_I v + n_E, \tag{3}$$

where δ_{SE} and δ_{IE} are estimation accuracy parameter for channel between S and E and that between I and E, respectively. The instantaneous SNR of the wiretap link can then be written as

$$\gamma_E = \frac{\left(\frac{P(1-\delta_{SE}^2)}{N_0 d_{SE}^\chi}\right)|h_{SE}|^2}{\frac{P(1-\delta_{IE}^2)}{N_0(1-\alpha)d_{IE}^\chi}|h_{IE}|^2 + \frac{P}{N_0(1-\alpha)d_{IE}^\chi}\delta_{IE} + \frac{P}{N_0 d_{SE}^\chi}\delta_{SE} + 1}. \tag{4}$$

FD Destination-Assisted Jamming

When D operates in FD mode, as illustrated in Fig. 5a, b, then it can use one antenna for reception and other for transmission. More specifically, there are two phases of transmission within a single time slot T. During the first phased, D harvests energy from S for αT amount of time. For the second phase, S transmits the information to D and D simultaneously receives the information and generates artificial noise.

Let S transmits the normalized energy signal s with transmission power P_S. Then, the received signal at D using (2) can be expressed as

$$\begin{aligned} y_D = & \sqrt{\frac{P(1-\delta_{SD}^2)}{d_{SD}^\chi}} h_{SD} s + \sqrt{\frac{P}{d_{SD}^\chi}}\delta_{SD} s v + \sqrt{\frac{P(1-\delta_{ID}^2)}{d_{ID}^\chi}} \\ & \times h_{ID} s_I + \sqrt{\frac{P}{d_{ID}^\chi}}\delta_{ID} s_I v + \sqrt{\frac{\zeta P\alpha(1-\delta_{SD}^2)(1-\delta_{DD}^2)}{(1-\alpha)d_{SD}^\chi}} \\ & \times h_{SD} h_{DD} s_{AN} + \sqrt{\frac{\zeta P\alpha(1-\delta_{SD}^2)}{(1-\alpha)d_{SD}^\chi}}\delta_{DD} h_{SD} s_{AN} v \\ & + \sqrt{\frac{\zeta P\alpha(1-\delta_{DD}^2)}{(1-\alpha)d_{SD}^\chi}}\delta_{SD} h_{DD} s_{AN} v + \sqrt{\frac{\zeta P\alpha}{(1-\alpha)d_{SD}^\chi}} \\ & \times \delta_{SD}\delta_{DD} s_{AN} v + n_D, \end{aligned} \tag{5}$$

where d_{SD} is the distance between S and D, h_{SD} is the estimated channel gain between S and D, and n_D is the AWGN with zero mean and N_0 variance. The harvested energy at D is represented as $E_1 = \frac{\zeta P|h_{SD}|\alpha T}{d_{SD}^\chi}$, where $0 \leq \zeta < 1$ is the energy conversion efficiency. Also, h_{DD} represents the loop-back interference channel coefficient. The instantaneous SNR at destination can then be represented as

$$\gamma_D = \frac{\left(\frac{P(1-\delta_{SD}^2)}{d_{SD}^\chi}\right)|h_{SD}|^2}{\left\{\frac{\zeta P\alpha(1-\delta_{SD}^2)(1-\delta_{DD}^2)}{(1-\alpha)d_{SD}^\chi}|h_{DD}|^2|h_{SD}|^2 + \frac{P(1-\delta_{ID}^2)}{d_{ID}^\chi}|h_{ID}|^2 + \frac{\zeta P\alpha(1-\delta_{SD}^2)}{(1-\alpha)d_{SD}^\chi} \times (\delta_{DD}|h_{SD}|^2 + \delta_{SD}|h_{DD}|^2) + \frac{\zeta P\alpha}{(1-\alpha)d_{SD}^\chi}\delta_{SD}\delta_{DD} + \frac{P}{d_{ID}^\chi}\delta_{ID} + \frac{P}{d_{SD}^\chi}\delta_{SD} + N_0\right\}}. \tag{6}$$

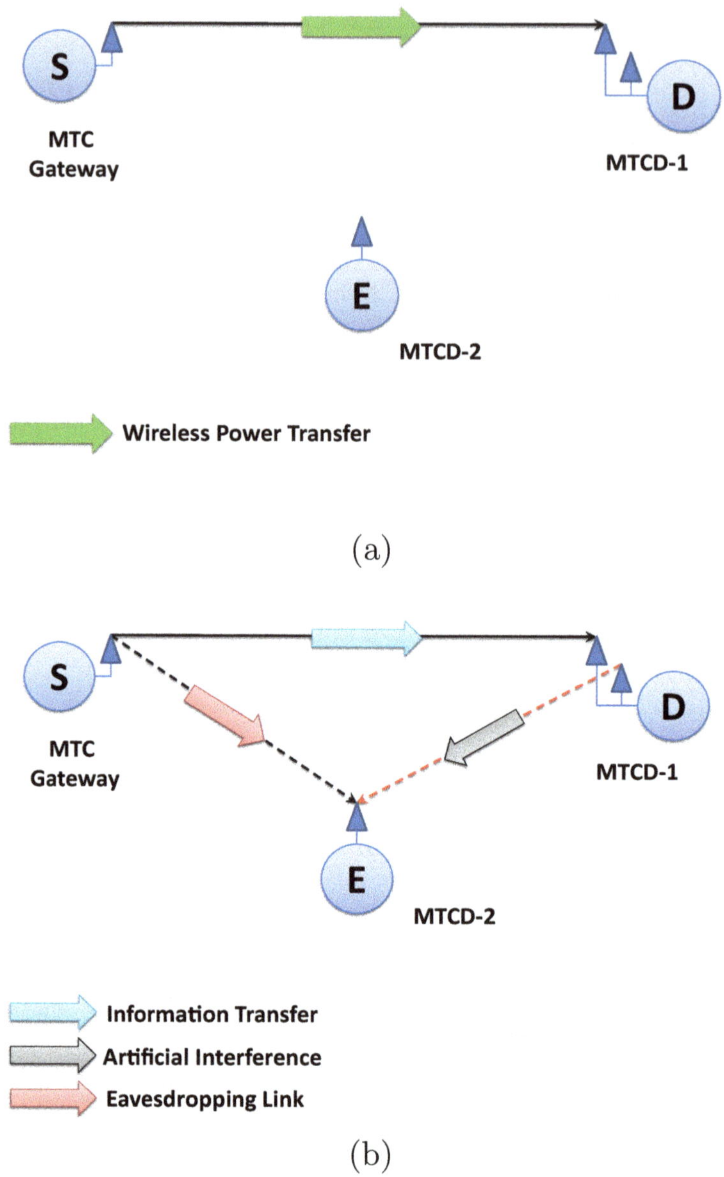

Fig. 5 System model for FD destination-assisted communication (**a**) Energy harvesting phase (**b**) Information transmission phase

After some straightforward simplifications, we obtain

$$\gamma_D = \frac{P(1-\delta_{SD}^2)|h_{SD}|^2(1-\alpha)d_{ID}^{\chi}}{\Big\{ d_{ID}^{\chi}\zeta P\alpha(1-\delta_{SD}^2)(1-\delta_{DD}^2)|h_{DD}|^2|h_{SD}|^2 + (1-\alpha)d_{SD}^{\chi}P(1-\delta_{ID}^2)|h_{ID}|^2}$$

$$+ d_{ID}^{\chi}\zeta P\alpha(1-\delta_{SD}^{2}) \times (\delta_{DD}|h_{SD}|^{2} + \delta_{SD}|h_{DD}|^{2}) + d_{ID}^{\chi}\zeta P\alpha\delta_{SD}\delta_{DD}$$

$$+ (1-\alpha)d_{SD}^{\chi}P\delta_{ID} + (1-\alpha)d_{ID}^{\chi}P\delta_{SD} + (1-\alpha)d_{SD}^{\chi}d_{ID}^{\chi}N_0\Big\}$$

$$= \frac{(1-\alpha)\Psi_1}{\alpha\Psi_2 + (1-\alpha)\Psi_3 + \alpha\Psi_4 + \alpha\Psi_5 + (1-\alpha)\Psi_6 + (1-\alpha)\Psi_7 + (1-\alpha)\Psi_8} \tag{7}$$

where $\Psi_1 = P(1-\delta_{SD}^{2})|h_{SD}|^{2}d_{ID}^{\chi}$, $\Psi_2 = d_{ID}^{\chi}\zeta P(1-\delta_{SD}^{2})(1-\delta_{DD}^{2})|h_{DD}|^{2}|h_{SD}|^{2}$, $\Psi_3 = d_{SD}^{\chi}P(1-\delta_{ID}^{2})|h_{ID}|^{2}$, $\Psi_4 = d_{ID}^{\chi}\zeta P(1-\delta_{SD}^{2})(\delta_{DD}|h_{SD}|^{2} + \delta_{SD}|h_{DD}|^{2})$, $\Psi_5 = d_{ID}^{\chi}\zeta P\delta_{SD}\delta_{DD}$, $\Psi_6 = d_{SD}^{\chi}P\delta_{ID}$, $\Psi_7 = d_{ID}^{\chi}P\delta_{SD}$, and $\Psi_8 = d_{SD}^{\chi}d_{ID}^{\chi}N_0$.

Using this power, D can send artificial noise to degrade the received signal at the eavesdropper E with power given as

$$P_D = \frac{E_1}{(1-\alpha)T} = \frac{\zeta P|h_{SD}|^{2}\alpha}{(1-\alpha)d_{SD}^{\chi}}, \tag{8}$$

Now the signal received at E, after receiving destination generated noise, can be written as

$$y_E = \sqrt{\frac{P(1-\delta_{SE}^{2})}{d_{SE}^{\chi}}}h_{SE}s + \sqrt{\frac{P}{d_{SE}^{\chi}}}\delta_{SE}sv + \sqrt{\frac{P(1-\delta_{IE}^{2})}{(1-\alpha)d_{IE}^{\chi}}}$$

$$\times h_{IE}s_I + \sqrt{\frac{P}{(1-\alpha)d_{IE}^{\chi}}}\delta_{IE}s_I v + \sqrt{\frac{\zeta P\alpha(1-\delta_{SD}^{2})(1-\delta_{DE}^{2})}{(1-\alpha)d_{SD}^{\chi}d_{DE}^{\chi}}}$$

$$\times h_{SD}h_{DE}s_{AN} + \sqrt{\frac{\zeta P\alpha(1-\delta_{SD}^{2})}{(1-\alpha)d_{SD}^{\chi}d_{DE}^{\chi}}}\delta_{DE}h_{SD}s_{AN}v$$

$$+ \sqrt{\frac{\zeta P\alpha(1-\delta_{DE}^{2})}{(1-\alpha)d_{SD}^{\chi}d_{DE}^{\chi}}}\delta_{SD}h_{DE}s_{AN}v + \sqrt{\frac{\zeta P\alpha}{(1-\alpha)d_{SD}^{\chi}d_{DE}^{\chi}}}$$

$$\times \delta_{SD}\delta_{DE}s_{AN}v + n_E. \tag{9}$$

Using (9), the instantaneous SINR at E can be written as

$$\gamma_E = \frac{\left(\frac{P(1-\delta_{SE}^{2})}{d_{SE}^{\chi}}\right)|h_{SE}|^{2}}{\Big\{\frac{\zeta P\alpha(1-\delta_{SD}^{2})(1-\delta_{DE}^{2})}{(1-\alpha)d_{SD}^{\chi}d_{DE}^{\chi}}|h_{DE}|^{2}|h_{SD}|^{2} + \frac{P(1-\delta_{IE}^{2})}{(1-\alpha)d_{IE}^{\chi}}|h_{IE}|^{2} + \frac{\zeta P\alpha(1-\delta_{SD}^{2})}{(1-\alpha)d_{SD}^{\chi}d_{DE}^{\chi}}(\delta_{DE}|h_{SD}|^{2}}$$

$$+ \delta_{SD}|h_{DE}|^{2}) + \frac{\zeta P\alpha}{(1-\alpha)d_{SD}^{\chi}d_{DE}^{\chi}}\delta_{SD}\delta_{DE} + \frac{P}{(1-\alpha)d_{IE}^{\chi}}\delta_{IE} + \frac{P}{d_{SE}^{\chi}}\delta_{SE} + N_0\Big\} \tag{10}$$

4.2 Secrecy Outage Probability Analysis

In this section, we will derive the expression of secrecy outage probability for both direct and FD mode. It is worth highlighting that secrecy performance analysis in previous works (see [65, 66, 69] and references therein) does not differentiate between secure and reliable communications. In other words, the outage event occurs when some information is leaked to the eavesdroppers or the receiver cannot successfully decode the transmission [72]. However, to fully understand the dynamics of secrecy outage probability in M2M networks, we analyze the secrecy performance under the condition of successful message transmission at the receiver. Thus, the secrecy outage probability can be defined as the likelihood of the event when the channel capacity from source-to-eavesdropper link becomes greater than the increment rate under nonadaptive encoding [72], and it can be represented as

$$P_{sop} = \Pr\{C_e > R_t - R_s | successful\ decoding\} \tag{11}$$

where $C_e = \log_2(1 + \gamma_e)$ is the channel capacity of the wiretap link, R_t is the rate of transmitted code word, and R_s is the rate of embedded information secrecy bits [73]. The difference between R_t and R_s actually determines the cost of the secrecy. In fact, $(R_t - R_s)$ is the information rate sacrificed to incorporate randomness in the signal to provide security against the eavesdropper.

Dedicated Jamming

As explained earlier, in direct transmission scheme, no jamming is performed by the destination node, and all the harvested energy is used for information decoding. Thus, we use (11) and (4) to find the expression of secrecy outage probability which is given as

$$P_{sop,d} = \Pr\left\{\log_2\left[1 + \frac{\left(\frac{P(1-\delta_{SE}^2)}{N_0 d_{SE}^{\chi}}\right)|h_{SE}|^2}{\frac{P(1-\delta_{IE}^2)}{N_0(1-\alpha)d_{IE}^{\chi}}|h_{IE}|^2 + \frac{P}{N_0(1-\alpha)d_{IE}^{\chi}}\delta_{IE} + \frac{P}{N_0 d_{SE}^{\chi}}\delta_{SE} + 1}\right] > R_t - R_s\right\} \tag{12}$$

Since Rayleigh fading is assumed between all the links, the power gains of the wireless channel will follow exponential distribution. Thus, we can simplify the above equation as

$$P_{sop,d} = 1 - \frac{\exp\left(-\frac{P\gamma_0}{N_0(1-\alpha)\lambda_{SE} d_{IE}^{\chi}}\delta_{IE} - \frac{P\gamma_0}{N_0 d_{SE}^{\chi}\lambda_{SE}}\delta_{SE} - \frac{\gamma_0}{\lambda_{SE}}\right)}{\lambda_{IE}}$$

$$\int_0^\infty \exp\left(-\frac{P(1-\delta_{IE}^2)\gamma_0 x}{N_0(1-\alpha)d_{IE}^\chi \lambda_{SE}} - \frac{x}{\lambda_{IE}}\right)dx, \tag{13}$$

where $\lambda_{SE} = \mathbb{E} = \{|h_{SE}|^2\}$, $\lambda_{IE} = \mathbb{E} = \{|h_{IE}|^2\}$ and $\gamma_0 = 2^{R_t - R_s} - 1$. After applying some straightforward mathematical steps, the expression in (13) resolves to

$$P_{sop,d} = 1 - \frac{N_0(1-\alpha)d_{IE}^\chi \lambda_{SE} \exp\left(-\frac{P\gamma_0}{N_0(1-\alpha)\lambda_{SE} d_{IE}^\chi}\delta_{IE} - \frac{P\gamma_0}{N_0 d_{SE}^\chi \lambda_{SE}}\delta_{SE} - \frac{\gamma_0}{\lambda_{SE}}\right)}{P\lambda_{IE}(1-\delta_{IE}^2)\gamma_0 + N_0(1-\alpha)d_{IE}^\chi \lambda_{IE}}. \tag{14}$$

FD Destination-Assisted Jamming

Note that the eavesdropper SNR expression in (10) is too complex and cannot be used to obtain a closed-form expression of secrecy outage probability. Instead, we use the approximate expression of eavesdropper SNR which using (10) can be written as

$$\gamma_E \approx \frac{\Psi_1 X_1}{\Psi_2 X_2 X_3 + \Psi_3 X_4 + \Psi_4 X_3 + \Psi_5 X_2 + \Theta}, \tag{15}$$

where $X_1 = |h_{SE}|^2$, $X_2 = |h_{DE}|^2$, $X_3 = |h_{SD}|^2$, $X_4 = |h_{IE}|^2$, $\Psi_1 = \frac{P(1-\delta_{SE}^2)}{N_0 d_{SE}^\chi}$, $\Psi_2 = \frac{\zeta P\alpha(1-\delta_{SD}^2)(1-\delta_{DE}^2)}{N_0(1-\alpha)d_{SD}^\chi d_{DE}^\chi}$, $\Psi_3 = \frac{P(1-\delta_{IE}^2)}{N_0(1-\alpha)d_{IE}^\chi}$, $\Psi_4 = \frac{\zeta P\alpha(1-\delta_{SD}^2)\delta_{DE}}{N_0(1-\alpha)d_{SD}^\chi d_{DE}^\chi}$, $\Psi_5 = \frac{\zeta P\alpha(1-\delta_{SD}^2)\delta_{SD}}{N_0(1-\alpha)d_{SD}^\chi d_{DE}^\chi}$, $\Theta = \frac{\zeta P\alpha}{(1-\alpha)d_{SD}^\chi d_{DE}^\chi}\delta_{SD}\delta_{DE} + \frac{P}{(1-\alpha)d_{IE}^\chi}\delta_{IE} + \frac{P}{d_{SE}^\chi}\delta_{SE} + 1$

Now, the secrecy outage probability for FD transmission scheme can be written as

$$\begin{aligned} P_{sop,fd} &= \Pr(\gamma_E > \gamma_0) = \Pr\left(\frac{\Psi_1 X_1}{\Psi_2 X_2 X_3 + \Psi_3 X_4 + \Psi_4 X_3 + \Psi_5 X_2 + \Theta} > \gamma_0\right) \\ &= 1 - \int_0^\infty \int_0^\infty \int_0^\infty \exp\left(-\frac{\Psi_2\gamma_0 x_2 x_3 + \Psi_3\gamma_0 x_4 + \Psi_4\gamma_0 x_3 + \Psi_5\gamma_0 x_2 + \Theta\gamma_0}{\Psi_1\lambda_1}\right) \\ &\quad \times \exp\left(-\frac{x_2}{\lambda_2}\right)\exp\left(-\frac{x_3}{\lambda_3}\right)\exp\left(-\frac{x_4}{\lambda_4}\right)dx_2 dx_3 dx_4 \end{aligned} \tag{16}$$

The above integrations can be simplified as

$$P_{sop,fd} = 1 - \exp\left(-\frac{\Theta\gamma_0}{\Psi_1\lambda_1}\right)\int_0^\infty \frac{\exp\left(-\frac{\Psi_4\gamma_0 x_3}{\Psi_1\lambda_1}\right)\lambda_4\lambda_2(\Psi_1\lambda_1)^2}{(\Psi_1\lambda_1 + \lambda_4\Psi_3\gamma_0)(\Psi_1\lambda_1 + \Psi_5\lambda_2 + \lambda_2\Psi_2 x_3)}dx_3 \tag{17}$$

With the help of [74, (8.21)], the above expression resolves to

$$P_{sop,fd} = 1 - \exp\left(-\frac{\Theta\gamma_0}{\Psi_1\lambda_1} + \left(\frac{\Psi_1\lambda_1}{\Psi_2\lambda_2} + \frac{\Psi_5}{\Psi_2}\right)\frac{\Psi_4\gamma_0}{\Psi_1\lambda_1}\right)\frac{\lambda_4(\Psi_1\lambda_1)^2}{\Psi_1\Psi_2\lambda_1 + \Psi_2\Psi_3\lambda_4\gamma_0}$$
$$\times \mathbb{E}i\left(\frac{\Psi_4\gamma_0}{\Psi_2\lambda_2} + \frac{\Psi_5\Psi_4\gamma_0}{\Psi_2\Psi_1\lambda_1}\right) \quad (18)$$

where $\mathbb{E}i(.)$ is the exponential integral function.

4.3 Results and Discussion

This section provides some numerical results on the analytical results derived in previous sections. Unless stated otherwise, we have used the following values of system parameters for generation of upcoming result: $P = -5$ dB, $\alpha = 0.4$, $\delta = \delta_{SD} = \delta_{ID} = \delta_{SE} = \delta_{DE} = 0.2$, $\chi = 2$, $zeta = 0.8$, $R_t = 2$, and $R_s = 1$ bit/sec/Hz.

In Fig. 6, we have plotted secrecy outage probability for increasing values of power conversion efficiency (ζ). Here, we have only shown the results for FD destination-assisted jamming as ζ has no impact on the dedicated jamming scheme. It can be observed that an increase in ζ results in decreasing the secrecy outage probability of the system. At lower values of increment rate (i.e., $R_t - R_s$), the

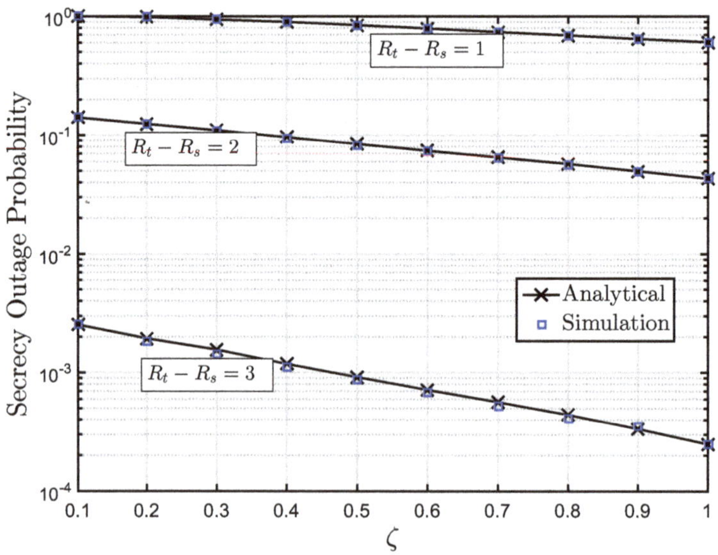

Fig. 6 Secrecy outage probability as a function of ζ

impact of ζ on secrecy outage probability is not much prominent. However, as the increment rate increases from 1 to 3, the secrecy outage probability decreases from threefolds for the same value of ζ. Additionally, the simulation results closely follow the analytical results which verify the analytical expressions derived in the above sections.

Figure 7 illustrates the effect of transmit power on the secrecy outage probability. In this figure, we have compared the secrecy performance of dedicated and FD jamming techniques. It can be noted that the dedicated jamming technique outperforms the FD jamming. However, this performance gain comes at a cost of introducing dedicated jamming node with additional power requirements. Moreover, we observe that the secrecy outage probability decreases with an increase in transmit power for the case of dedicated jamming, but the same is not true for FD jamming. Specifically, the secrecy outage probability for FD jamming technique first decreases and then becomes constant at higher values of transmit power. This is mainly due to the self-interference caused by the jamming antenna. At higher powers, the effect of self-interference becomes more prominent, hence introducing a secrecy outage floor.

In Fig. 8, we have shown secrecy outage probability as a function of increment rate. For comparative purposes, we have plotted the analytical and simulation curves for both dedicated and FD jamming techniques. It can be seen that the dedicated jamming technique again outperforms the FD jamming. Moreover, at lower values of increment rate, the difference between secrecy performance of FD and dedicated jamming techniques is nominal. However, as the increment rate increases, the

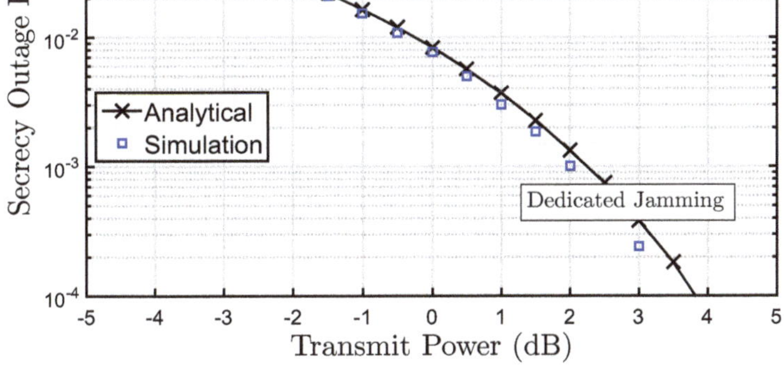

Fig. 7 Secrecy outage probability against increasing transmit power

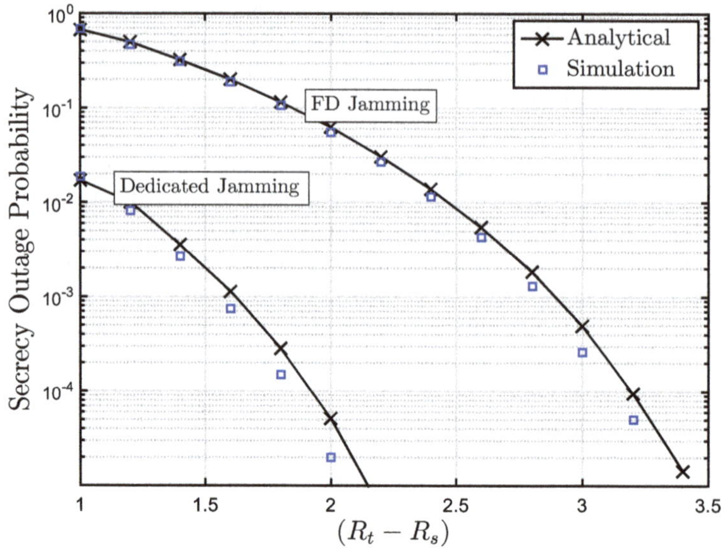

Fig. 8 Secrecy outage probability against $R_t - R_s$

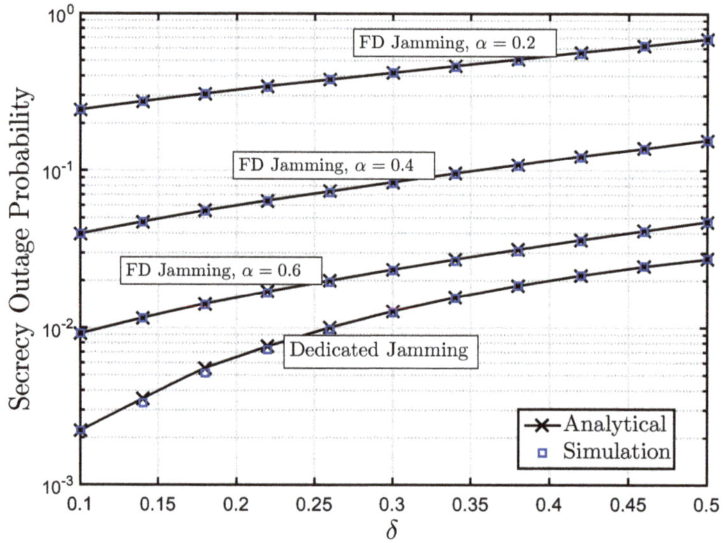

Fig. 9 Secrecy outage probability versus δ

difference between the two techniques becomes more vivid. Still, for both cases, an increase in $R_t - R_s$ results in decreasing the secrecy outage probability.

Figure 9 plots the secrecy outage probability against increasing values of channel estimation accuracy factor δ. It can be observed that an increase in the value of

δ results in increasing the secrecy outage probability for both dedicated and FD jamming. It is because the channel estimation errors increase resulting in less accurate knowledge of the channels. Again, the dedicated jamming outperforms the FD jamming. However, interestingly, as the values of α increases from 0.2 to 0.6, the secrecy performance of FD jamming becomes approaches that of dedicated jamming technique. It is because more portion of time is being used for harvesting energy which is then used for performing jamming in the second phase.

4.4 Conclusions

This chapter starts with an introduction to M2M communications, its applications, and communication requirements. The chapter also provides a brief overview of energy harvesting techniques with special focus on RF energy harvesting. This is followed by a detailed discussion on recent trends in the domain of PLS and recently introduced secure energy harvesting protocols. By building on this knowledge, we analyzed the secrecy performance of an M2M communication system under the availability of imperfect CSI. We then compared two jamming techniques, namely, FD jamming and dedicated jamming. We have shown that the dedicated jamming outperforms FD jamming at the cost of introducing an additional node with additional power requirements. We have also shown that by increasing the duration of energy harvesting, the FD destination-assisted jamming can also perform very close to dedicated jamming.

4.5 Future Research Directions

M2M communication promises to assist IoTs to provide seamless connectivity between devices of future networks. Exploiting M2M communication between MTCDs in close proximity presents many advantages. But existing techniques for M2M networks would not be able to meet the extreme capacity demands of future networks. Discussions of new standards are underway in the academia and industry in order to envision the requirements of M2M networks. There is no doubt that M2M communications have a wider range of use cases and have the ability to support existing and emerging technologies to meet the increase in demand for data rates. However, the same would not be possible unless the issues of energy and security are addressed.

Since M2M communication manages information exchange between two different machines without control (or with limited control) of the humans, therefore, new procedures need to be created with a specific end goal to improve range, vitality, and effectiveness of M2M networks. Moreover, there is a need to lessen the communication interval in cell-based M2M networks. One way to lessen the communication gap is through cooperative M2M communications. Cooperative relaying is a better technique to cope with noise enhancement, but trusting a relay may cause information leakage. In other words, cooperative communication can compromise the security of transmitted messages due to the involvement of untrusted intermediate relays. An MTCD has to forward information to another MTCD without knowledge of its network state. Furthermore, the handover of information can take place often over multiple hops which can result in compromising both the authenticity and the privacy of the data. Thus, there is a need to design efficient PLS techniques for M2M communications which can ensure secrecy over multiple hops.

In addition, PLS-aided M2M communication is yet to see improvements owing to the fact that many research domains are new, and research done is limited in several different aspects. PLS-aided M2M networks can also apply relay selection to have superior secrecy performance. As it is obvious from the discussion, the area of PLS is still being explored, and a number of variations and improvements are possible for each discussed area, while a number of new areas need to be explored to make sure that security is provided in the worst possible eavesdropping cases.

References

1. V. Vijayaraghavan, R. Agarwal, Security and privacy across connected environments, in *Connected Environments for the Internet of Things* (Springer, Cham, 2017), pp. 19–39
2. A. Elmangoush, A. Al-Hezmi, T. Magedanz, The development of M2M standards for ubiquitous sensing service layer, in *Globecom Workshops (GC Wkshps), 2014* (IEEE, 2014), pp. 624–629
3. M. Chen, J. Wan, S. González-Valenzuela, X. Liao, V.C. Leung, A survey of recent developments in home M2M networks. IEEE Commun. Surv. Tutorials **16**(1), 98–114 (2014)
4. J. Rico, B. Cendn, J. Lanza, J. Valio, Bringing IoT to hospital logistics systems demonstrating the concept, in *2012 IEEE Wireless Communications and Networking Conference Workshops (WCNCW)*, 2012, pp. 196–201
5. A.S. Lalos, L. Alonso, C. Verikoukis, Model based compressed sensing reconstruction algorithms for ECG telemonitoring in WBANs. Digital Signal Process. **35**, 105–116 (2014)
6. S. Subhani, H. Shi, J.F.G. Cobben, A survey of technical challenges in wireless machine-to-machine communication for smart grids, in *2015 50th International Universities Power Engineering Conference (UPEC)*, 2015, pp. 1–6
7. M. Alsabaan, W. Alasmary, A. Albasir, K. Naik, Vehicular networks for a greener environment: a survey. IEEE Commun. Surv. Tutorials **15**(3), 1372–1388 (2013)
8. Y. Zeng, N. Xiong, L.T. Yang, Y. Zhang, Cross-layer routing in wireless sensor networks for machine-to-machine intelligent hazard monitoring applications, in *2011 IEEE Conference on Computer Communications Workshops (INFOCOM WKSHPS)*, 2011, pp. 206–211
9. A. Laya, L. Alonso, J. Alonso-Zarate, Is the random access channel of LTE and LTE-A suitable for M2M communications? A survey of alternatives. IEEE Commun. Surv. Tutorials **16**(1), 4–16 (2014)
10. E. Ronen, A. Shamir, A.-O. Weingarten, C. O'Flynn, IoT goes nuclear: creating a ZigBee chain reaction, in *2017 IEEE Symposium on Security and Privacy (SP)* (IEEE, 2017), pp. 195–212
11. F. Jameel, M.A. Javed, D.N.K. Jayakody, S.A. Hassan, On secrecy performance of industrial Internet of things. Internet Technol. Lett. **1**(2), e32. https://doi.org/10.1002/itl2.32
12. F. Jameel, F. Khan, M.A.A. Haider, A.U. Haq, Secrecy analysis of relay assisted device-to-device systems under channel uncertainty, in *International Conference on Frontiers of Information Technology (FIT)*, 2017, pp. 345–349

13. F. Jameel, S. Wyne, I. Krikidis, Secrecy outage for wireless sensor networks. IEEE Commun. Lett. **21**(7), 1565–1568 (2017)
14. I. Ganchev, M. Curado, A. Kassler, *Wireless Networking for Moving Objects: Protocols, Architectures, Tools, Services and Applications*, vol. 8611 (Springer, Cham, 2014)
15. R. Lu, X. Li, X. Liang, X. Shen, X. Lin, GRS: The green, reliability, and security of emerging machine to machine communications. IEEE Commun. Mag. **49**(4), 28–35 (2011)
16. S. Kitagami, Y. Kaneko, T. Suganuma, Method of autonomic load balancing for long polling in M2M service system, in *26th International Conference on Advanced Information Networking and Applications Workshops (WAINA)* (IEEE, 2012), pp. 294–299
17. M. Saedy, V. Mojtahed, Ad Hoc M2M communications and security based on 4G cellular system, in *Wireless Telecommunications Symposium (WTS)* (IEEE, 2011), pp. 1–5
18. Z. Fu, X. Jing, S. Sun, Application-based identity management in M2M system. IET Conf. Proc. 211–215 (2011)
19. C. Hongsong, F. Zhongchuan, Z. Dongyan, Security and trust research in M2M system, in *IEEE International Conference on Vehicular Electronics and Safety (ICVES)* (IEEE, 2011), pp. 286–290
20. T.D.P. Perera, D.N.K. Jayakody, S.K. Sharma, S. Chatzinotas, J. Li, Simultaneous wireless information and power transfer (SWIPT): recent advances and future challenges. IEEE Commun. Surv. Tutorials **20**(1), 264–302 (2018)
21. D.N.K. Jayakody, J. Thompson, S. Chatzinotas, S. Durrani, *Wireless Information and Power Transfer: A New Paradigm for Green Communications* (Springer, Cham, 2017)
22. M. Raju, M. Grazier, Energy harvesting. ULP meets energy harvesting: a game-changing combination for design engineers. http://focus.ti.com/lit/wp/slyy018/slyy018.pdf (2008) 3413–3423
23. W.K. Seah, Z.A. Eu, H.-P. Tan, Wireless sensor networks powered by ambient energy harvesting (WSN-HEAP)-survey and challenges, in *Wireless Communication, Vehicular Technology, Information Theory and Aerospace & Electronic Systems Technology, 2009. 1st International Conference on Wireless VITAE 2009* (IEEE, 2009), pp. 1–5
24. B.H. Calhoun, D.C. Daly, N. Verma, D.F. Finchelstein, D.D. Wentzloff, A. Wang, S.-H. Cho, A.P. Chandrakasan, Design considerations for ultra-low energy wireless microsensor nodes. IEEE Trans. Comput. **54**(6), 727–740 (2005)
25. C. Alippi, C. Galperti, An adaptive system for optimal solar energy harvesting in wireless sensor network nodes. IEEE Trans. Circuits Syst. I Regul. Pap. **55**(6), 1742–1750 (2008)
26. L. Mateu, C. Codrea, N. Lucas, M. Pollak, P. Spies, Human body energy harvesting thermogenerator for sensing applications, in *2007 International Conference on Sensor Technologies and Applications (SENSORCOMM 2007)*, 2007, pp. 366–372
27. F. Mansourkiaie, L.S. Ismail, T.M. Elfouly, M.H. Ahmed, Maximizing lifetime in wireless sensor network for structural health monitoring with and without energy harvesting. IEEE Access **5**, 2383–2395 (2017)
28. M. Najimi, A. Ebrahimzadeh, S.M.H. Andargoli, A. Fallahi, Lifetime maximization in cognitive sensor networks based on the node selection. IEEE Sens. J. **14**(7), 2376–2383 (2014)
29. H. Salarian, K.W. Chin, F. Naghdy, An energy-efficient mobile-sink path selection strategy for wireless sensor networks. IEEE Trans. Veh. Technol. **63**(5), 2407–2419 (2014)
30. Y. Chen, Q. Zhao, On the lifetime of wireless sensor networks. IEEE Commun. Lett. **9**(11), 976–978 (2005)
31. J.W. Jung, M.A. Weitnauer, On using cooperative routing for lifetime optimization of multi-hop wireless sensor networks: analysis and guidelines. IEEE Trans. Commun. **61**(8), 3413–3423 (2013)
32. F. Jameel, S. Wyne, Secrecy outage of SWIPT in the presence of cooperating eavesdroppers. AEU Int. J. Electron. Commun. **77**, 23–26 (2017)
33. K. Huang, V.K. Lau, Enabling wireless power transfer in cellular networks: architecture, modeling and deployment. IEEE Trans. Wirel. Commun. **13**(2), 902–912 (2014)
34. L. Liu, R. Zhang, K.-C. Chua, Multi-antenna wireless powered communication with energy beamforming. IEEE Trans. Commun. **62**(12), 4349–4361 (2014)

35. L.R. Varshney, Transporting information and energy simultaneously, in *IEEE International Symposium on Information Theory, 2008. ISIT 2008* (IEEE, 2008), pp. 1612–1616
36. F. Jameel, Z. Hamid, F. Jabeen, S. Zeadally, M.A. Javed, A Survey of device-to-device communications: research issues and challenges. IEEE Commun. Surv. Tutorials 1–1 (2018). https://doi.org/10.1109/COMST.2018.2828120
37. P. Grover, A. Sahai, Shannon meets Tesla: wireless information and power transfer, in *2010 IEEE International Symposium on Information Theory Proceedings (ISIT)* (IEEE, 2010), pp. 2363–2367
38. D.W.K. Ng, E.S. Lo, R. Schober, Wireless information and power transfer: energy efficiency optimization in OFDMA systems. IEEE Trans. Wirel. Commun. **12**(12), 6352–6370 (2013)
39. Q. Shi, L. Liu, W. Xu, R. Zhang, Joint transmit beamforming and receive power splitting for MISO SWIPT systems. IEEE Trans. Wirel. Commun. **13**(6), 3269–3280 (2014)
40. D. Li, C. Shen, Z. Qiu, Two-way relay beamforming for sum-rate maximization and energy harvesting, in *2013 IEEE International Conference on Communications (ICC)* (IEEE, 2013), pp. 3115–3120
41. D.S. Michalopoulos, H.A. Suraweera, R. Schober, Simultaneous information transmission and wireless energy transfer via selecting one out of two relays, in *2014 6th International Symposium on Communications, Control and Signal Processing (ISCCSP)* (IEEE, 2014), pp. 318–321
42. F. Jameel, S. Wyne, G. Kaddoum, T.Q. Duong, A comprehensive survey on cooperative relaying and jamming strategies for physical layer security. IEEE Commun. Surv. Tutorials 1–1 (2018). https://doi.org/10.1109/COMST.2018.2865607
43. F. Jameel, M. Faisal, A.A. Haider, A.A. Butt, Physical layer security under Rayleigh/Weibull and Hoyt/Weibull fading, in *2017 13th International Conference on Emerging Technologies (ICET)*, 2017, pp. 1–5. https://doi.org/10.1109/ICET.2017.8281715
44. A. Zhang, L. Wang, X. Ye, X. Lin, Light-weight and robust security-aware D2D-assist data transmission protocol for mobile-health systems. IEEE Trans. Inf. Forensics Secur. **12**(3), 662–675 (2017)
45. G. Verma, P. Yu, B.M. Sadler, Physical layer authentication via fingerprint embedding using software-defined radios. IEEE Access **3**, 81–88 (2015)
46. L.Y. Paul, G. Verma, B.M. Sadler, Wireless physical layer authentication via fingerprint embedding. IEEE Commun. Mag. **53**(6), 48–53 (2015)
47. L. Shi, M. Li, S. Yu, J. Yuan, BANA: body area network authentication exploiting channel characteristics. IEEE J. Sel. Areas Commun. **31**(9), 1803–1816 (2013)
48. W. Hou, X. Wang, J.-Y. Chouinard, A. Refaey, Physical layer authentication for mobile systems with time-varying carrier frequency offsets. IEEE Trans. Commun. **62**(5), 1658–1667 (2014)
49. K.M. Borle, B. Chen, W.K. Du, Physical layer spectrum usage authentication in cognitive radio: analysis and implementation. IEEE Trans. Inf. Forensics Secur. **10**(10), 2225–2235 (2015)
50. L. Wang, H. Wu, G.L. Stüber, Cooperative jamming-aided secrecy enhancement in P2P communications with social interaction constraints. IEEE Trans. Veh. Technol. **66**(2), 1144–1158 (2017)
51. W. Wang, K.C. Teh, K.H. Li, Enhanced physical layer security in D2D spectrum sharing networks. IEEE Wirel. Commun. Lett. **6**(1), 106–109 (2017)
52. Z. Shu, Y. Qian, S. Ci, On physical layer security for cognitive radio networks. IEEE Netw. **27**(3), 28–33 (2013)
53. L. Liu, R. Zhang, K.-C. Chua, Secrecy wireless information and power transfer with miso beamforming, in *Global Communications Conference (GLOBECOM), 2013 IEEE* (IEEE, 2013), pp. 1831–1836
54. F. Jameel, D.N.K. Jayakody, M.F. Flanagan, C. Tellambura, Secure communication for separated and integrated receiver architectures in SWIPT, in *IEEE Wireless Communications and Networking Conference (WCNC)*, 2018, pp. 1–6
55. I. Krikidis, S. Timotheou, S. Nikolaou, G. Zheng, D.W.K. Ng, R. Schober, Simultaneous wireless information and power transfer in modern communication systems. IEEE Commun. Mag. **52**(11), 104–110 (2014)

56. H. Ju, R. Zhang, Optimal resource allocation in full-duplex wireless-powered communication network. IEEE Trans. Commun. **62**(10), 3528–3540 (2014)
57. O. Taghizadeh, A. Zamani, R. Mathar, Physical-layer security for simultaneous information and power transfer in full-duplex multi-user networks, in *Proceedings of the 20th International ITG Workshop on Smart Antennas (WSA 2016)* (VDE, 2016), pp. 1–8
58. Y. Bi, H. Chen, Accumulate and jam: towards secure communication via a wireless-powered full-duplex jammer. IEEE J. Sel. Top. Sign. Proces. **10**(8), 1538–1550 (2016)
59. W. Wu, B. Wang, Z. Deng, H. Zhang, Secure beamforming for full-duplex wireless powered communication systems with self-energy recycling. IEEE Wirel. Commun. Lett. **6**(2), 146–149 (2017)
60. B. Yang, W. Wang, B. Yao, Q. Yin, Destination assisted secret wireless communication with cooperative helpers. IEEE Signal Process Lett. **20**(11), 1030–1033 (2013)
61. Y. Liu, A.P. Petropulu, H.V. Poor, Joint decode-and-forward and jamming for wireless physical layer security with destination assistance, in *Conference Record of the Forty Fifth Asilomar Conference on Signals, Systems and Computers (ASILOMAR)* (IEEE, 2011), pp. 109–113
62. Y. Liu, A.P. Petropulu, Relay selection and scaling law in destination assisted physical layer secrecy systems, in *Statistical Signal Processing Workshop (SSP), 2012 IEEE* (IEEE, 2012), pp. 381–384
63. Y. Liu, J. Li, A.P. Petropulu, Destination assisted cooperative jamming for wireless physical-layer security. IEEE Trans. Inf. Forensics Secur. **8**, 682–694 (2013)
64. S. Luo, J. Li, A.P. Petropulu, Uncoordinated cooperative jamming for secret communications. IEEE Trans. Inf. Forensics Secur. **8**(7), 1081–1090 (2013)
65. S. Luo, J. Li, A. Petropulu, Outage constrained secrecy rate maximization using cooperative jamming, in *Statistical Signal Processing Workshop (SSP), 2012 IEEE* (IEEE, 2012), pp. 389–392
66. W. Aman, G.A.S. Sidhu, T. Jabeen, F. Gao, S. Jin, Enhancing physical layer security in dual-hop multiuser transmission, in *Wireless Communications and Networking Conference (WCNC)* (IEEE, 2016), pp. 1–6
67. Z. Ding, Z. Ma, P. Fan, Asymptotic studies for the impact of antenna selection on secure two-way relaying communications with artificial noise. IEEE Trans. Wirel. Commun. **13**(4), 2189–2203 (2014)
68. Y. Tang, J. Xiong, D. Ma, X. Zhang, Robust artificial noise aided transmit design for MISO wiretap channels with channel uncertainty. IEEE Commun. Lett. **17**(11), 2096–2099 (2013)
69. W. Li, M. Ghogho, B. Chen, C. Xiong, Secure communication via sending artificial noise by the receiver: outage secrecy capacity/region analysis. IEEE Commun. Lett. **16**(10), 1628–1631 (2012)
70. S.H. Tsai, H.V. Poor, Power allocation for artificial-noise secure MIMO precoding systems. IEEE Trans. Signal Process. **62**(13), 3479–3493 (2014)
71. C. Zhong, H.A. Suraweera, G. Zheng, I. Krikidis, Z. Zhang, Wireless information and power transfer with full duplex relaying. IEEE Trans. Commun. **62**(10), 3447–3461 (2014)
72. X. Zhou, M.R. McKay, B. Maham, A. Hjorungnes, Rethinking the secrecy outage formulation: a secure transmission design perspective. IEEE Commun. Lett. **15**(3), 302–304 (2011)
73. X. Zhou, Y. Zhang, L. Song, *Physical Layer Security in Wireless Communications* (CRC Press, Baton Rouge, 2016)
74. I.S. Gradshteyn, I.M. Ryzhik, *Table of Integrals, Series, and Products* (Academic, Amsterdam, 2014)

Furqan Jameel received his BS in Electrical Engineering (under ICT R&D funded Program) in 2013 from the Lahore Campus of COMSATS Institute of Information Technology (CIIT), Pakistan. In 2017, he received his Masters degree in Electrical Engineering (funded by prestigious Higher Education Commission Scholarship) at the Islamabad Campus of CIIT. In September 2018, he visited Simula Research Laboratory and the University of Oslo, Norway. Currently, he is a first-year Ph.D. student at the University of Jyvaskyla, Finland where his research interests include modeling and performance enhancement of vehicular networks, physical layer security, ambient backscatter communications, and wireless power transfer. He is the recipient of outstanding reviewer award 2017 from Elsevier.

Muhammad Awais Javed is currently working as an Assistant Professor at COMSATS University Islamabad, Pakistan. He completed his Ph.D. in Electrical Engineering from The University of Newcastle, Australia in Feb. 2015 and B.Sc. in Electrical Engineering from University of Engineering and Technology Lahore, Pakistan in Aug. 2008. From July 2015-June 2016, he worked as a Postdoc Research Scientist at Qatar Mobility Innovations Center (QMIC) on SafeITS project.

Dushantha Nalin K. Jayakody received his Ph.D. in Electronics, Electrical, and Communications Engineering in 2013 from the University College Dublin, Ireland. He received his M.Sc. in Electronics and Communications Engineering from the Department of Electrical and Electronics Engineering, Eastern Mediterranean University, Turkey (under the University full graduate scholarship) and ranked as the first merit position holder of the department, and B.E. in Electronics Engineering (with first-class honors) from Pakistan and was ranked as the merit position holder of the University (under SAARC Scholarship). From 2014 to 2016, he was a Postdoc Research Fellow at the Institute of Computer Science, University of Tartu, Estonia, and Department of Informatics, University of Bergen, Norway. From 2016, he is a Professor at the School of Computer Science & Robotics, National Research Tomsk Polytechnic University, Russia, where he also serves as the Director of Tomsk Infocomm Lab. As of 2019, he also serves as the Director of School of Graduate Studies at the Sri Lanka Technological Campus (SLTC), Sri Lanka. Dr. Jayakody has received the best paper award from the IEEE International Conference on Communication, Management and Information Technology (ICCMIT) in 2017. Dr. Jayakody has published over 100 international peer-reviewed journal and conference papers, and he is an IEEE senior member. His research interests include PHY layer prospective of 5G communications, cooperative wireless communications, device-to-device communications, LDPC codes, unmanned aerial vehicle, etc. Dr. Jayakody is a member of IEEE and he has served as workshop chair, session chair, or technical program committee member for various international conferences, such as IEEE PIMRC 2013/2014, IEEE WCNC 2014-2018, IEEE VTC 2015-2018, etc. He currently serves as an area editor of journals such as *Physical Communications* (Elsevier), *Information* (MDPI), and *Internet Technology Letters* (Wiley) and is on the advisory board of the journal *Sci* (MDPI). Also, he serves as a reviewer for various *IEEE Transactions* and other journals.

Beam-Domain Full-Duplex Massive MIMO Transmission in the Cellular System

Kui Xu, Xiaochen Xia, Yurong Wang, Wei Xie, and Dongmei Zhang

1 Introduction

The ever-growing challenges for significant traffic growth driven by mobile Internet and Internet of Things have made system capacity enhancement one of the most important features in next-generation wireless communication systems. The general consensus is that the aggregate data rate will increase by roughly 1000 by 2020. Massive multiple-input multiple-output (MIMO), which is first advocated in [1], is identified as one of the key enabling technologies to achieve this goal due to its strong potential in boosting the spectral efficiency (SE) of wireless networks [1, 2]. The term massive MIMO indicates that the base station (BS) employs a number of antennas (typically several tens to hundreds) much larger than the number of active data streams per time-frequency resource. The benefits of massive MIMO are twofold. First, massive MIMO produces a large surplus of degrees of freedom, which can be used to create asymptotically orthogonal channels and deliver near interference-free signals for each user equipment (UE). In this way, the network SE is enhanced significantly because more UEs can be served in parallel and each UE suffers from less interference. On the other hand, the tremendous array gain of the large-scale antenna array also helps to save transmit power and thus potentially improves the energy efficiency.

Massive MIMO was originally designed for time-division duplex (TDD) system [1–7], since by exploiting the channel reciprocity in TDD setting, the required channel state information (CSI) for downlink transmission at the BS can be easily obtained via uplink training [1]. The training overhead scales linearly with the number of user equipments (UEs) and is independent with the number of BS antennas, which is acceptable in most of the typical scenarios. As frequency-division

K. Xu (✉) · X. Xia · Y. Wang · W. Xie · D. Zhang
Army Engineering University of PLA, Nanjing, China

D. N. K. Jayakody et al. (eds.), *5G Enabled Secure Wireless Networks*,
https://doi.org/10.1007/978-3-030-03508-2_6

duplex (FDD) dominates the current wireless cellular systems, the application of massive MIMO in FDD system is even more desirable. In FDD massive MIMO, the downlink training and corresponding CSI feedback yield an unacceptably high overhead, which poses a significant bottleneck on the achievable SE. One attempt of practical FDD massive MIMO is called joint spatial division and multiplexing (JSDM) [8], where the correlation between channels is exploited to reduce the training and feedback dimensions. Another scheme that enables FDD massive MIMO is called beam division multiple access (BDMA) [9]. The BDMA gets rid of the need of CSI at transmitter and provides strong potential to realize massive MIMO gain in FDD system. Moreover, other innovative approaches, such as the phase-only beamforming [10] and two-stage beamforming [11], are also promising solutions to the FDD massive MIMO.

In TDD and FDD massive MIMO systems (namely, half-duplex (HD) massive MIMO systems), the uplink and downlink UEs must be allocated with orthogonal time slots or frequency bands, which results in insufficient utilization of time-frequency resources. Inspired by the recent development of full-duplex (FD) communication [17], co-time co-frequency uplink and downlink (CCUD) transmission becomes another option in the cellular system. Although attractive in SE, CCUD transmission is considered challenging due to the strong self-interference (SI) caused by the signal leakage between BS transmitter and receiver, especially when the BS is equipped with large-scale antenna arrays. In the small-scale MIMO system, the SI can be mitigated by the active SI cancellation (SIC) scheme, such as digital/circuit domain SIC and spatial suppression [17]. However, the impractical requirement of instantaneous high-dimension SI channel knowledge makes these technologies difficult when applied in the large-scale antenna system. The passive SIC can be applied in the SI channel-unware environment, but it fails to provide satisfactory SIC level when used alone [17]. On the other hand, to support the CCUD transmission, the BS employs a separate antenna configuration[1] where two separate large-scale antenna arrays are used for transmission and reception, respectively [18]. In this case, the downlink channel reciprocity is commonly considered as unavailable [19]. Without reciprocity, the training overhead to obtain the downlink CSI scales linearly with the number of BS antennas, which poses another big challenge. In [18], to make the system feasible, the authors assumed that each transmit antenna of BS is also connected with a receive radio-frequency chain so that it can receive the pilot signal. In this case, the downlink reciprocity can still be exploited, however, at cost of additional hardware complexity.

Note that the CCUD transmission in the cellular system with massive MIMO BS has been investigated recently in several works (see [12–16] and the references therein). For example, the authors in [12–14] studied the SE performance of

[1] We mention that there exists another choice of shared antenna configuration which uses a single antenna array for transmission and reception. However, under the current technologies, the shared configuration is still difficult in the multi-antenna system due to the significant cross talk between different antennas [20]. Therefore, it will not be considered in this chapter.

CCUD transmission in both macro-cell and small-cell environments. The linear beamforming design of the BS for CCUD transmission has been considered in [15] and [16]. However, most of these works are based on the assumption that the SI has been suppressed to a reasonable level and the uplink/downlink channel can be efficiently obtained. As a result, the aforementioned challenges are still not fully addressed.

In this chapter, we investigate the feasibility of CCUD transmission in the cellular system with massive MIMO BS. The contributions are summarized as follows.

1. By exploiting the beam-domain representation of channels based on the basis expansion model [23], we prove that massive MIMO channel matrix (vector) can be represented by a low-dimension effective beam-domain channel matrix (vector). Based on this property, we propose a beam-domain full-duplex (BDFD) massive MIMO scheme (BDFD scheme for short) to enable CCUD transmission in the cellular system. We show that the BDFD scheme achieves significant saving in uplink/downlink training and achieves the uplink and downlink sum capacities simultaneously as the number of BS antennas approaches to infinity.
2. Then, we investigate several important components for the practical implementation of BDFD scheme in the cellular system, including UEs grouping, effective beam-domain channel estimation, beam-domain data transmission, and interference control between uplink and downlink.
3. Finally, we examine the SE of BDFD scheme using the third-generation partnership project long-term evolution (3GPP LTE) simulation model for macro-cell environment. The results demonstrate the superiority of BDFD scheme over the TDD/FDD massive MIMO.

The rest of the paper is organized as follows. The system and channel models are described in Sects. 2, 3, and 4 considering the basic ideal and practical implementation of BDFD scheme, respectively. Section 5 presents the simulation results. Section 6 draws the conclusions.

Notation $\mathbb{E}(\cdot)$ denotes the expectation. $\delta(\cdot)$ denotes the Dirac delta function. $\mathbf{A}^{\{B_1,B_2\}}$ denotes the submatrix of $\mathbf{A}$ by keeping its rows indexed by set B_1 and columns indexed by set B_2. $\mathbf{A}^{\{B,:\}}$ ($\mathbf{A}^{\{:,B\}}$) denotes the submatrix of $\mathbf{A}$ by keeping its rows (columns) indexed by set B. $(\cdot)^T$, $(\cdot)^*$, $(\cdot)^H$, $|\cdot|$, $\|\cdot\|$, and $tr(\cdot)$ denote transpose, conjugate, conjugate-transpose, determinant, Frobenius norm, and trace of a matrix, respectively. $\mathbf{A}\succeq 0$ means that $\mathbf{A}$ is Hermitian positive semi-definite matrix. The frequently used symbols in this paper are summarized in Table 1.

2 System and Channel Models

Consider a single-cell system with a FD BS, a number of uplink UEs, and a number of downlink UEs as shown in Fig. 1a. We assume that all UEs are HD and have single antenna. To support the CCUD transmission, the BS employs two separate large-scale antenna arrays for transmission and reception, respectively. The uniform

Table 1 Summary of frequently used symbols

Symbols	Descriptions
$\mathbf{h}_{k_u}\left(\mathbf{h}_{k_d}\right)$	The channel vector between uplink UE k_u (downlink UE k_d) and BS
$\mathbf{H}_{SI}$	The SI channel matrix from transmit antenna array to receive antenna array of the BS
$\mathbf{H}_{g_u}\left(\mathbf{H}_{g_d}\right)$	The channel matrix between uplink group g_u (downlink group g_d) and BS
N	The number of transmit/receive antennas of BS
M_u,M_d,M_{SI}	The numbers of scattering clusters for uplink, downlink, and SI channels
$\left[\theta_{k_u,i}^{\min}, \theta_{k_u,i}^{\max}\right]$	The DOA region of uplink signal from UE k_u resulting from the ith scattering cluster
$\left[\theta_{k_d,i}^{\min}, \theta_{k_d,i}^{\max}\right]$	The DOD region of downlink signal to UE k_d resulting from the ith scattering cluster
$S_{\omega,i}(\cdot),\omega \in \{k_u, k_d, SI\}$	The product of the large-scale fading and channel power angle spectrum
$g_{u,k}(g_{d,k})$	The kth UE in the uplink group g_u (downlink group g_d)
$B_{k_u}\left(B_{k_d}\right)$	The active beam set of uplink group UE k_u (downlink group UE k_d)
$B_{g_u}\left(B_{g_d}\right)$	The active beam set of uplink group g_u (downlink group g_d)
$B_{SI,R}$,$B_{SI,T}$	The active beam sets of SI channel
$G_u(G_d)$	The set of all uplink (downlink) group
$K_{g_u}\left(K_{g_d}\right)$	The number of UEs in uplink group g_u (downlink group g_d)
$b_u(b_d)$	The number beams in uplink (downlink) UE group
σ	The variance of AWGN

linear arrays are assumed. In the practical implementation, the transmit and receive antenna arrays of the BS can be deployed on the opposite sides of a building with distance of tens of meters to reduce the SI.

We use $\mathbf{h}_{k_u} \in \mathbb{C}^{N\times 1}$ to denote the channel vector from the uplink UE k_u to the receive antenna array of BS and use $\mathbf{h}_{k_d} \in \mathbb{C}^{N\times 1}$ to denote the channel vector from the transmit antenna array of BS to downlink UE k_d, where N denotes the number of transmit/receive antennas at the BS.[2] We use $\mathbf{H}_{SI} \in \mathbb{C}^{N \times N}$ to denote the SI channel matrix from transmit antenna array to receive antenna array of the BS.

We consider the general cluster-based channel model [21] where the received signal at the BS from the uplink UE k_u is a sum of the contributions from M_u scattering clusters. The direction of arrival (DOA) of signals resulting from the ith cluster is within the region $\left[\theta_{k_u,i}^{\min}, \theta_{k_u,i}^{\max}\right]$. Thus, the channel vector between the uplink UE k_u and the BS can be expressed as [21]

$$\mathbf{h}_{k_u} = \sum_{i=1}^{M_u} \int_{\theta_{k_u,i}^{\min}}^{\theta_{k_u,i}^{\max}} \mathbf{a}(\theta)\, r_{k_u,i}(\theta)\, d\theta \tag{1}$$

[2]To simplify the notation, we assume the symmetric antenna deployment at the BS. Extension to the situation with different numbers of transmit and receive antennas is straightforward.

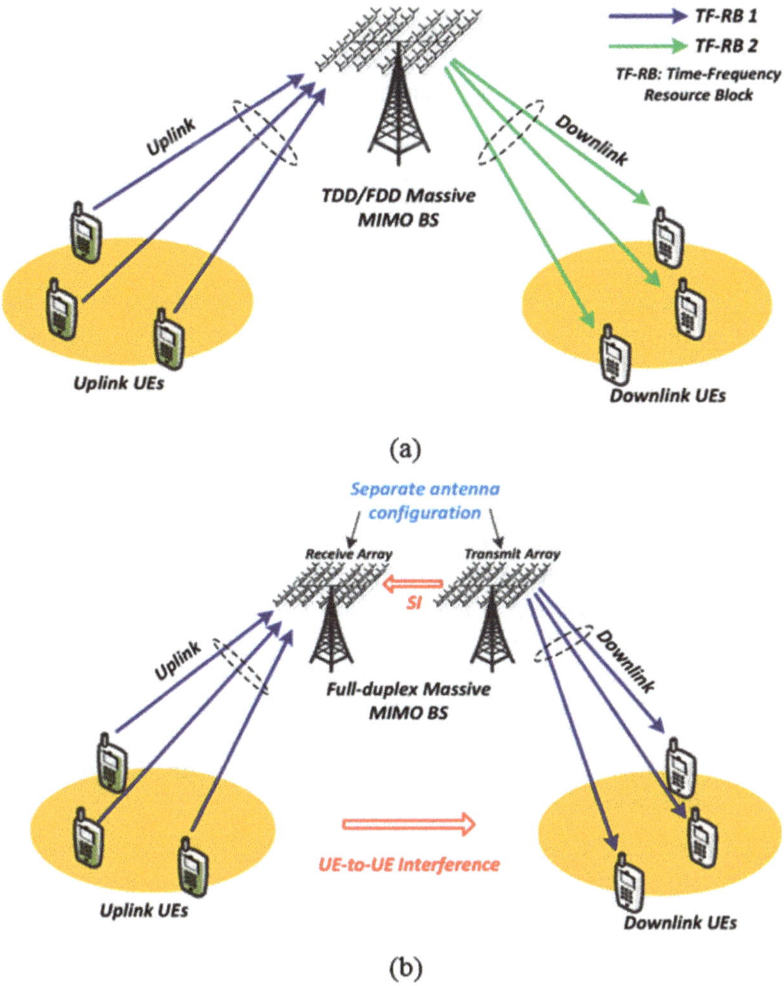

Fig. 1 TDD/FDD massive MIMO and FD massive MIMO systems. (**a**) TDD/FDD massive MIMO. (**b**) Full-duplex massive MIMO

where $\mathbf{a}(\theta) = [1, \exp(j2\pi d \sin(\theta)/\lambda), \cdots, \exp(j2\pi d(N-1)\sin(\theta)/\lambda)]^T$ is the array response vector with d and λ denoting the antenna spacing and carrier wavelength, respectively. $r_{k_u,i}(\theta)$ denotes the complex-valued response gain. In the above model, the DOA regions of signals from different scattering clusters are disjoint (otherwise, these signals should be considered from the same scattering cluster). Therefore, the number of scattering clusters is finite because $\sum_{i=1}^{M_u}\left(\theta_{k_u,i}^{\max} - \theta_{k_u,i}^{\min}\right) \leq 2\pi$.

Similarly, let $\left[\theta_{k_d,i}^{\min}, \theta_{k_d,i}^{\max}\right]$ be the direction of departure (DOD) region of signals resulting from the ith scattering clusters and let $r_{k_d,i}(\theta)$ denote the associated complex-valued response gain; the channel vector from the BS to downlink UE k_d can be written as

$$\mathbf{h}_{k_d} = \sum_{i=1}^{M_u} \int_{\theta_{k_d,i}^{\min}}^{\theta_{k_d,i}^{\max}} \mathbf{a}(\theta)\, r_{k_d,i}(\theta)\, d\theta \tag{2}$$

The SI signal can be viewed as the contributions of signals from M_{SI} scattering clusters with different DOA and DOD regions. Thus, the SI channel matrix $\mathbf{H}_{SI}$ can be expressed as

$$\mathbf{H}_{SI} = \sum_{i=1}^{M_{SI}} \int_{\theta_{R,i}^{\min}}^{\theta_{R,i}^{\max}} \int_{\theta_{T,i}^{\min}}^{\theta_{T,i}^{\max}} r_{SI,i}(\theta_R, \theta_T)\, \mathbf{a}(\theta_R)\, \mathbf{a}^H(\theta_T)\, d\theta_R d\theta_T \tag{3}$$

where $r_{SI,i}(\theta_R, \theta_T)$ denotes the complex-valued response gain. In real systems, the BS is commonly elevated at a relatively high altitude, e.g., on the top of a high building or a dedicated tower, so that there are few surrounding scatterers [24]. Moreover, we assume that the passive SIC scheme for infrastructure nodes in [25] has been used, and the direct path between transmit and receive antenna arrays of BS is virtually cancelled. Therefore, in this chapter, we assume that the number of scattering clusters for SI channel is small.

In (1), (2), and (3), the complex-valued response gains with different incidence angles are uncorrelated [22], that is,

$$\mathbb{E}\left[r_{k_u,i}(\theta)\, r_{k_u,i}^*(\theta)\right] = S_{k_u,i}(\theta)\, \delta\left(\theta - \theta'\right)$$

$$\mathbb{E}\left[r_{k_d,i}(\theta)\, r_{k_d,i}^*(\theta)\right] = S_{k_d,i}(\theta)\, \delta\left(\theta - \theta'\right)$$

$$\mathbb{E}\left[r_{SI,i}(\theta_R, \theta_T)\, r_{SI,i}^*(\theta_R, \theta_T)\right] = S_{SI,i}(\theta_R, \theta_T)\, \delta\left(\theta_R - \theta_R'\right) \delta\left(\theta_T - \theta_T'\right) \tag{4}$$

where $S_{\omega,i}(\cdot), \omega \in \{k_u, k_d, SI\}$ represents the product of the large-scale fading and channel power angle spectrum. Note that the considered model can be easily transformed into several well-known massive MIMO channel models. For example, by setting $M_u = M_d = 1$, we obtain the "one-ring" model studied in [8]. The "one-ring" model is typically used in the macro-cell environment where the uplink/downlink received signals are resulted from the scattering process in the vicinity of the UEs [21]. Moreover, by setting

$$r_{k_u,i}(\theta) = \sum_j r_{k_u,i,j}\, \delta\left(\theta - \theta_j\right)$$

$$r_{k_u,i}(\theta) = \sum_j r_{k_u,i,j}\delta(\theta - \theta_j)$$

$$r_{SI,i}(\theta_R, \theta_T) = \sum_{j,l} r_{k_{SI},i,j,l}\delta(\theta_R - \theta_{R,j})\,\delta(\theta_T - \theta_{T,j}) \tag{5}$$

we arrive at the ray-cluster-based spatial channel model which is usually used for millimeter wave MIMO systems [26]. Therefore, the results in this chapter can be readily applied in these scenarios.

3 Beam-Domain Full-Duplex Transmission Scheme

In this section, we propose a BDFD scheme to realize CCUD transmission in the cellular system. Using the basis expansion model, we first derive the beam-domain channel representation which is the projection of channel vector (matrix) on a common basis. The benefit of the beam-domain representation is that the channel becomes compressible in the beam domain under certain basis. Using this property, channel dimension required to be estimated can be greatly reduced. Moreover, by exploiting the structure of SI channel in the beam domain, it is possible to eliminate the SI without using the instantaneous SI channel knowledge and hence realize efficient CCUD transmission.

3.1 Beam-Domain Channel Representation

Under the basis expansion model [23], the uplink channel vector can be expanded from a set of uniform basis vectors $\{\mathbf{f}_1, \mathbf{f}_2, \cdots, \mathbf{f}_N\} \in \mathbb{C}^{N \times 1}$, that is

$$\mathbf{h}_{k_u} = \sum_{m=1}^{N} \tilde{h}_{k_u,m}\mathbf{f}_m = \mathbf{F}\tilde{\mathbf{h}}_{k_u} \tag{6}$$

where $\mathbf{F} = [\mathbf{f}_1, \mathbf{f}_2 \cdots, \mathbf{f}_N]$. Following [9], the basis vector $\mathbf{f}_i$ is also called a beam, and $\tilde{\mathbf{h}}_{k_u} = \left[\tilde{h}_{k_u,1}, \tilde{h}_{k_u,2}, \cdots, \tilde{h}_{k_u,N}\right]^T$ is called the beam-domain channel.[3] According to (1) and (6), we have

[3]Note that the idea of beam-domain channel was also studied in [9] for FDD massive MIMO system. However, we investigate the beam-domain properties of a more general channel model and present new capacity achieving scheme using these properties.

$$\tilde{\mathbf{h}}_{k_u} = \mathbf{F}^H \mathbf{h}_{k_u} = \sum_{i=1}^{M_u} \int_{\theta_{k_u,i}^{\min}}^{\theta_{k_u,i}^{\max}} \mathbf{F}^H \mathbf{a}(\theta)\, r_{k_u,i}(\theta)\, d\theta \tag{7}$$

To investigate the compressibility of beam-domain channel, we propose the following lemma.

Lemma 1 Consider the basis $\mathbf{F} = [\mathbf{f}_1, \mathbf{f}_2 \cdots, \mathbf{f}_N]$ with $\mathbf{f}_n = \frac{1}{\sqrt{N}}\left[1, \exp(j2\pi d\theta_n/\lambda), \cdots, \exp(j2\pi d(N-1)\theta_n/\lambda)\right]^T$. θ_n is defined as the beam angle of nth beam $\mathbf{f}_n$, which is selected so that the different beams are orthogonal. As the number of BS antennas N tends to infinity, the average beam-domain channel gain for the uplink UE k_u associated with the nth beam $\mathbf{f}_n$, i.e., $\mathbb{E}\left[\left|\tilde{h}_{k_u,n}\right|^2\right]$, has non-negligible value only when $\theta_n \in \cup_{i=1}^{M_u}\left[\sin\theta_{k_u,i}^{\min} - \varepsilon, \sin\theta_{k_u,i}^{\max} + \varepsilon\right]$, where $\varepsilon \geq 0$ and $\lim_{N\to\infty}\varepsilon = 0$.

Proof Using (7) and [28, Eq. (5)], $\mathbb{E}\left[\left|\tilde{h}_{k_u,n}\right|^2\right]$ can be written as

$$\begin{aligned}
\mathbb{E}\left[\left|\tilde{h}_{k_u,n}\right|^2\right] &= \sum_{i=1}^{M_u} \int_{\theta_{k_u,i}^{\min}}^{\theta_{k_u,i}^{\max}} \left|\mathbf{f}_n^H \mathbf{a}(\theta)\right|^2 S_{k_u,i}(\theta)\, d\theta \\
&= \sum_{i=1}^{M_u} \underbrace{N \int_{\theta_{k_u,i}^{\min}}^{\theta_{k_u,i}^{\max}} \mathrm{asinc}_N^2\left(\frac{d}{\lambda}\theta_n - \frac{d}{\lambda}\sin\theta\right) S_{k_u,i}(\theta)\, d\theta}_{Y_i}
\end{aligned} \tag{8}$$

where $\mathrm{asinc}_N(x)$ is the aliased sinc function, which is defined as $\mathrm{asinc}_N(x) = \sin(N\pi x)/(N\sin(\pi x))$. The envelope of the squared aliased sinc function is shown in Fig. 2. Assuming $\theta_n = \sin\theta_{k_u,i}^{\max} + \varepsilon$ with $\varepsilon > 0$ and $i \in \{1, 2, \cdots, M_u\}$, the ith term in the summation of (8), named as Y_i, can be upper bounded as

$$\begin{aligned}
Y_i &\leq N \max_{\theta\in\left[\theta_{k_u,i}^{\min}, \theta_{k_u,i}^{\max}\right]} S_{k_u,i}(\theta) \int_{\theta_{k_u,i}^{\min}}^{\theta_{k_u,i}^{\max}} \mathrm{asinc}_N^2\left(\tfrac{d}{\lambda}\theta_n - \tfrac{d}{\lambda}\sin\theta\right) d\theta \\
&\leq N \max_{\theta\in\left[\theta_{k_u,i}^{\min}, \theta_{k_u,i}^{\max}\right]} S_{k_u,i}(\theta) \sum_{l=L_i^{\min}}^{L_i^{\max}} \int_{\vartheta_{l+1}}^{\vartheta_l} \mathrm{asinc}_N^2\left(\tfrac{d}{\lambda}\theta_n - \tfrac{d}{\lambda}\sin\theta\right) d\theta \\
&\leq N\left(L_i^{\max} - L_i^{\min}\right) \max_{\theta\in\left[\theta_{k_u,i}^{\min}, \theta_{k_u,i}^{\max}\right]} S_{k_u,i}(\theta) \int_{\vartheta_{L_i^{\min}+1}}^{\vartheta_{L_i^{\min}}} \mathrm{asinc}_N^2\left(\tfrac{d}{\lambda}\theta_n - \tfrac{d}{\lambda}\sin\theta\right) d\theta \\
&\leq \frac{\left(L_i^{\max} - L_i^{\min}\right)\left(\vartheta_{L_i^{\min}} - \vartheta_{L_i^{\min}+1}\right)}{N\left(\pi L_i^{\min}\right)^2} \max_{\theta\in\left[\theta_{k_u,i}^{\min}, \theta_{k_u,i}^{\max}\right]} S_{k_u,i}(\theta)
\end{aligned} \tag{9}$$

where $L_i^{\min} = \left\lceil Nd\left(\theta_n - \sin\theta_{k_u,i}^{\max}\right)/\lambda\right\rceil$ and $L_i^{\max} = \left\lfloor Nd\left(\theta_n - \sin\theta_{k_u,i}^{\min}\right)/\lambda\right\rfloor$. $\vartheta_l = \arcsin(\theta_n - \lambda l/dN)$ is the lth zero point of the function $\mathrm{asinc}_N^2(d(\theta_n - \sin\theta)/\lambda)$.

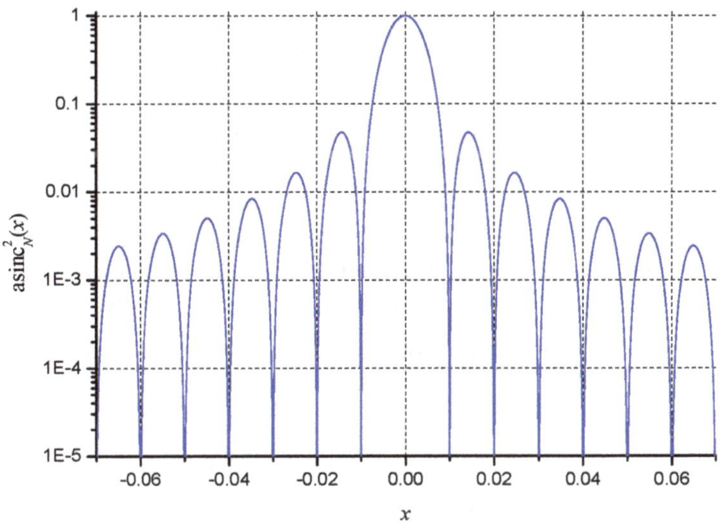

Fig. 2 Envelope of the squared aliased sinc function, where $N = 100$

The lth term in the summation of the second step is the integral over the lth side lobe of the squared aliased sinc function as shown in Fig. 2. The third step is based on the fact that the power of the side lobe of aliased sinc function is a decreasing function of its index. The last step is obtained by using the property that the aliased sinc function converges to the standard sinc function as $N \to \infty$ [29, Ch. 3] and using the fact $\sin x \le 1$.

Note that $L_i^{\max} - L_i^{\min}$ scales with $\mathcal{O}(N)$. Moreover, as $N \to \infty$, $\vartheta_{L_i^{\min}} - \vartheta_{L_i^{\min}+1}$ can be replaced by the differential of the arcsin$x$ at point $x = \theta_n - \lambda L_i^{\min}/dN$, i.e.,

$$\begin{aligned} \vartheta_{L_i^{\min}} - \vartheta_{L_i^{\min}+1} &= \frac{\lambda}{dN} \left.\frac{\mathrm{d}(\arcsin x)}{\mathrm{d}x}\right|_{x=\theta_n - \frac{\lambda L_i^{\min}}{dN}} \\ &= \frac{\lambda}{dN}\left(1 - \left(\theta_n - \frac{\lambda L_i^{\min}}{dN}\right)^2\right)^{-1/2} = \mathcal{O}\left(\frac{1}{N}\right) \end{aligned} \tag{10}$$

Recall that we have assumed $\theta_n = \sin\theta_{k_u,i}^{\max} + \varepsilon$. Thus, we have $L_i^{\min} = \lfloor dN\varepsilon/\lambda \rfloor$. Based on the results in the above, the upper bound of Y_i given by the last step of (9) converges to zero when $N \to \infty$ as long as $\varepsilon \ge \mathcal{O}(N^u)$ with $u > -3/2$. If we can choose $\varepsilon = \mathcal{O}\left(N^{-1}\right)$, then ε approaches to 0 if $N \to \infty$. In the same way, we can obtain the similar result if $\theta_n = \sin\theta_{k_u,i}^{\max} - \varepsilon$. Therefore, as $N \to \infty$, Y_i ($\forall i \in \{1, 2, \cdots, M_u\}$) has non-negligible value only if $\theta_n \in \left[\sin\theta_{k_u,i}^{\min} - \varepsilon, \sin\theta_{k_u,i}^{\max} + \varepsilon\right]$, where $\lim_{N\to\infty}\varepsilon = 0$. This completes the proof.

Note that the upper bound of Y_i in (9) is generally not tight. However, this does not impact the analysis because $Y_i \overset{N\to\infty}{\to} 0$ as long as its upper bound converges to 0.

If we consider $\sin\theta$ as the virtual DOA of the uplink signal, in Lemma 1 we actually select the beam angle to mimic virtual DOA. That is why only the beam-domain channel elements with beam angles within $\cup_{i=1}^{M_u}\left[\sin\theta_{k_u,i}^{\min}-\varepsilon, \sin\theta_{k_u,i}^{\max}+\varepsilon\right]$ have non-negligible gains. From Lemma 1, the beam-domain channel vector exhibits the desired compressibility in the large N regime when the considered basis is used. Therefore, the basis in Lemma 1 will be employed in the following.

Example 1 As a concrete example of the compressibility, we consider a scenario with $N = 128$, $M_u = 1$, and $\left[\theta_{k_u,1}^{\min}, \theta_{k_u,1}^{\max}\right] = \left[24.3^\circ, 35.7^\circ\right]$. This corresponds to "one-ring" model with about 30 m scattering radius and 300 m BS-to-UE distance [21]. The normalized average beam-domain channel gain, which is defined as $\mathbb{E}\left[\left|\tilde{h}_{k_u,n}\right|^2\right] / \max_{n'=1,2,\cdots,N} \mathbb{E}\left[\left|\tilde{h}_{k_u,n'}\right|^2\right]$, is plotted in Fig. 3. The beam-domain channel elements whose beam angles are within $\left[\sin\theta_{k_u,1}^{\min}, \sin\theta_{k_u,1}^{\max}\right]$ are marked in red. From the figure, we can see that the gains of these channel elements are much higher than the remaining, which matches well with the results in Lemma 1. In fact, from the simulation results, about 96.6% of channel power is captured by these channel elements (less than 10% of the all elements) when $N = 128$. This value becomes 96.9% and 98.1% if we increase N to 256 and 512, respectively.

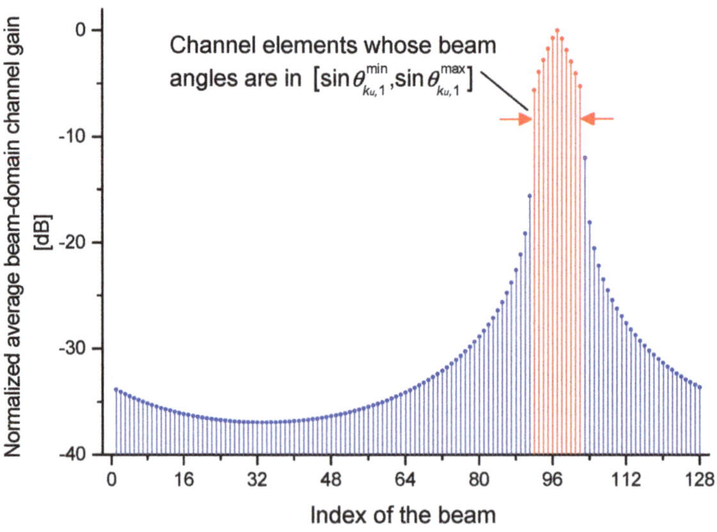

Fig. 3 Normalized average beam-domain channel gain as a function of index of the associated beam, where $\theta_n = \frac{\lambda}{d}\left(\frac{n}{N}-\frac{1}{2}\right)$, and $d = \lambda/2$

Based on Lemma 1, we can approximate the channel vector from uplink UE k_u to BS as

$$\mathbf{h}_{k_u} \approx \sum_{m \in B_{k_u}} \tilde{h}_{k_u,m} \mathbf{f}_m = \mathbf{F}^{\{:,B_{k_u}\}} \tilde{\mathbf{h}}_{k_u}^{\{B_{k_u},:\}} \tag{11}$$

where B_{k_u} is called the active beam set which contains the indexes of beams with non-negligible beam-domain channel gains. $\mathbf{F}^{\{:,B_{k_u}\}}$ is called the active beam space, whose columns are consisted of the beams in B_{k_u}. The reduced-dimension beam-domain channel vector $\tilde{\mathbf{h}}_{k_u}^{\{B_{k_u},:\}} \in \mathbf{C}^{|B_{k_u}|\times 1}$ is called the effective beam-domain channel. Note that (11) holds with equality as $N \to \infty$ according to Lemma 1. Based on (1) and (11), the effective beam-domain channel vector can be expressed as

$$\tilde{\mathbf{h}}_{k_u}^{\{B_{k_u},:\}} = \left(\mathbf{F}^{\{:,B_{k_u}\}}\right)^H \mathbf{h}_{k_u} = \sum_{i=1}^{M_u} \int_{\theta_{k_u,i}^{\min}}^{\theta_{k_u,i}^{\max}} \left(\mathbf{F}^{\{:,B_{k_u}\}}\right)^H \mathbf{a}(\theta)\, r_{k_u,i}(\theta)\, d\theta \tag{12}$$

From (11), the original channel vector can be recovered from the effective beam-domain channel vector if the DOA (and hence the active beam set) information is known. As a result, in order to obtain $\mathbf{h}_{k_u}$, it is enough to estimate the $|B_{k_u}|$-dimension effective beam-domain channel during the training phase. This can potentially result in significant saving of the training resource.

In the practical scenario with arbitrary finite number of BS antennas, the active beam set B_{k_u} can be obtained by solving a cardinality minimization problem, with constraint that most of the channel power is captured by the effective beam-domain channel vector, i.e.,

$$\begin{aligned} &\min_{B_{k_u}} |B_{k_u}| \\ &\text{s.t.} \begin{cases} \dfrac{\mathbb{E}\left[\left\|\tilde{\mathbf{h}}_{k_u}^{\{B_{k_u},:\}}\right\|^2\right]}{\mathbb{E}\left[\left\|\tilde{\mathbf{h}}_{k_u}\right\|^2\right]} = \dfrac{\sum_{i=1}^{M_u} \int_{\theta_{k_u,i}^{\min}}^{\theta_{k_u,i}^{\max}} \left\|\left(\mathbf{F}^{\{:,B_{k_u}\}}\right)^H \mathbf{a}(\theta)\right\|^2 S_{k_u,i}(\theta) d\theta}{\sum_{i=1}^{M_u} \int_{\theta_{k_u,i}^{\min}}^{\theta_{k_u,i}^{\max}} \|\mathbf{F}^H \mathbf{a}(\theta)\|^2 S_{k_u,i}(\theta) d\theta} \geq \eta \\ \eta < 1 \end{cases} \end{aligned} \tag{13}$$

where $\eta < 1$ denotes the threshold and should be chosen closed to 1 in real implementation. The problem is combinatorial and difficult to solve in closed form. However, since (13) is related only with DOA information which is slow time-varying,[4] we can build off-line table of B_{k_u} for different DOA regions to reduce the computation load. If N is large enough, to further reduce the complexity, we can

[4]Since the DOA/DOD information is slow time-varying, we assume that these parameters can be obtained perfectly at the BS through the long-term estimation [27].

simply select the beams whose angles are in $\theta_m \in \cup_{i=1}^{M_{SI}} \left[\sin\theta_{R,i}^{\min} - \varepsilon, \sin\theta_{R,i}^{\max} + \varepsilon\right]$ to constitute B_{k_u}. According to Lemma 1 (also demonstrated by in Example 1), the beam-domain channel elements associated with these beams contain almost all the channel power in the large N regime.

Similarly, with the basis expansion model, the channel vector from the BS to downlink UE k_d and the associated effective beam-domain channel vector can be expressed as

$$\tilde{\mathbf{H}}_{SI}^{\{B_{SI,R},B_{SI,T}\}} = \left(\mathbf{F}^{\{:,B_{SI,R}\}}\right)^H \mathbf{H}_{SI}\mathbf{F}^{\{:,B_{SI,T}\}}$$
$$= \sum_{i=1}^{M_{SI}} \int_{\theta_{R,i}^{\min}}^{\theta_{R,i}^{\max}} \int_{\theta_{T,i}^{\min}}^{\theta_{T,i}^{\max}} r_{SI,i}\left(\theta_R, \theta_T\right) \left(\mathbf{F}^{\{:,B_{SI,R}\}}\right)^H \mathbf{a}\left(\theta_R\right)\mathbf{a}^H\left(\theta_T\right)\mathbf{F}^{\{:,B_{SI,T}\}} d\theta_R d\theta_T \tag{14}$$

where the active beam set B_{k_d} can be design by solving a similar problem as that in (13).

To exploit the compressibility of SI channel, the basis expansion is performed for column and row spaces of $\mathbf{H}_{SI}$ simultaneously, which results in $\mathbf{H}_{SI} = \mathbf{F}\tilde{\mathbf{H}}_{SI}\mathbf{F}^H$. As a generalization of Lemma 1, we have the following Lemma.

Lemma 2 The average beam-domain SI channel gain $\mathbb{E}\left[\left|\left[\tilde{\mathbf{H}}_{SI}\right]_{m,n}\right|^2\right] = \mathbb{E}\left[\left|\mathbf{f}_m^H\mathbf{H}_{SI}\mathbf{f}_n\right|^2\right]$ has non-negligible value only when the beam angle of $\mathbf{f}_m$ lies in $\cup_{i=1}^{M_{SI}} \left[\sin\theta_{R,i}^{\min} - \varepsilon, \sin\theta_{R,i}^{\max} + \varepsilon\right]$, and meanwhile the beam angle of $\mathbf{f}_n$ lies in $\cup_{i=1}^{M_{SI}} \left[\sin\theta_{T,i}^{\min} - \varepsilon, \sin\theta_{T,i}^{\max} + \varepsilon\right]$, where ε approaches to zero in the large N regime.

Lemma 2 can be simply proved by using a similar procedure at that in the proof of Lemma 1. Therefore, the detailed proof is omitted due to space limitation. From Lemma 2, the SI channel matrix can be approximated as $\mathbf{H}_{SI} \approx \mathbf{F}^{\{:,B_{SI,R}\}}\tilde{\mathbf{H}}_{SI}^{\{B_{SI,R},B_{SI,T}\}}\left(\mathbf{F}^{\{:,B_{SI,T}\}}\right)^H$, where the effective beam-domain SI channel matrix $\tilde{\mathbf{H}}_{SI}^{\{B_{SI,R},B_{SI,T}\}} \in \mathbf{C}^{|B_{SI,R}|\times|B_{SI,T}|}$ can be expressed as

$$\tilde{\mathbf{H}}_{SI}^{\{B_{SI,R},B_{SI,T}\}} = \left(\mathbf{F}^{\{:,B_{SI,R}\}}\right)^H \mathbf{H}_{SI}\mathbf{F}^{\{:,B_{SI,T}\}}$$
$$= \sum_{i=1}^{M_{SI}} \int_{\theta_{R,i}^{\min}}^{\theta_{R,i}^{\max}} \int_{\theta_{T,i}^{\min}}^{\theta_{T,i}^{\max}} r_{SI,i}\left(\theta_R, \theta_T\right) \left(\mathbf{F}^{\{:,B_{SI,R}\}}\right)^H \mathbf{a}\left(\theta_R\right)\mathbf{a}^H\left(\theta_T\right)\mathbf{F}^{\{:,B_{SI,T}\}} d\theta_R d\theta_T \tag{15}$$

The active beam sets $B_{SI,R}$ and $B_{SI,R}$ can be determined by solving the problem (16) on the top of the next page, where $\tilde{\mathbf{H}}_{SI}$ is defined as $\tilde{\mathbf{H}}_{SI} = \mathbf{F}^H\mathbf{H}_{SI}\mathbf{F}$ and $\eta < 1$ denotes the threshold in (16), shown at the bottom of the page.

$$\min_{\mathcal{B}_{SI,R},\mathcal{B}_{SI,T}} \max\left\{\left|\mathcal{B}_{SI,R}\right|,\left|\mathcal{B}_{SI,T}\right|\right\} \text{ s.t.} \frac{\mathbb{E}\left[\left\|\tilde{\mathbf{H}}_{SI}^{\{\mathcal{B}_{SI,R},\mathcal{B}_{SI,T}\}}\right\|^2\right]}{\mathbb{E}\left[\left\|\tilde{\mathbf{H}}_{SI}\right\|^2\right]}$$
$$= \frac{\sum_{i=1}^{M_{SI}} \int_{\theta_{R,i}^{\min}}^{\theta_{R,i}^{\max}} \int_{\theta_{T,i}^{\min}}^{\theta_{T,i}^{\max}} S_{SI,i}(\theta_R,\theta_T)\left\|\left(\mathbf{F}^{\{:,\mathcal{B}_{SI,R}\}}\right)^H \mathbf{a}(\theta_R)\mathbf{a}^H(\theta_T)\mathbf{F}^{\{:,\mathcal{B}_{SI,T}\}}\right\|^2 d\theta_R d\theta_T}{\sum_{i=1}^{M_{SI}} \int_{\theta_{R,i}^{\min}}^{\theta_{R,i}^{\max}} \int_{\theta_{T,i}^{\min}}^{\theta_{T,i}^{\max}} S_{SI,i}(\theta_R,\theta_T)\left\|\mathbf{F}^H\mathbf{a}(\theta_R)\mathbf{a}^H(\theta_T)\mathbf{F}\right\|^2 d\theta_R d\theta_T} \geq \eta \tag{16}$$

3.2 Beam-Domain Full-Duplex Transmission

The key idea of the BDFD scheme lies in partitioning UEs according to their active beam sets to realize efficient CCUD transmission. In particular, we divide the UEs into groups according the following two criteria (UE grouping criteria):

1. Criterion 1: The uplink/downlink UEs with the same active beam set are collected in the same group. The active beam sets of different uplink/downlink groups are non-overlapping. Mathematically, letting $\mathcal{B}_{g_u}$ and $\mathcal{B}_{g_u'}$ be the active beam sets of two arbitrary uplink groups g_u and g_u', and letting $\mathcal{B}_{g_d}$ and $\mathcal{B}_{g_d'}$ be the active beam sets of two arbitrary downlink groups g_d and g_d', we have $\mathcal{B}_{g_u} \cap \mathcal{B}_{g_u'} = \varnothing$ and $\mathcal{B}_{g_d} \cap \mathcal{B}_{g_d'} = \varnothing$.
2. Criterion 2: Let $\mathcal{G}_u$ and $\mathcal{G}_d$ be the sets of uplink UE groups and downlink UE groups, respectively. The active beam sets $\mathcal{B}_{g_u}$ and $\mathcal{B}_{g_d}$ satisfy $\left(\cup_{g_u\in\mathcal{G}_u}\mathcal{B}_{g_u}\right) \cap \mathcal{B}_{SI,R} = \varnothing$ or $\left(\cup_{g_d\in\mathcal{G}_d}\mathcal{B}_{g_d}\right) \cap \mathcal{B}_{SI,T} = \varnothing$.

For the sake of illustration, in the following we assume that each active beam set contains the same number of beams, i.e., $\left|\mathcal{B}_{g_u}\right| = b_u$ and $\left|\mathcal{B}_{g_d}\right| = b_d$. Define the index $g_{u,k} = k + \sum_{g_u'=1}^{g_u-1} K_{g_u'}$ to denote the kth uplink UE of the uplink group g_u, where K_{g_u} is the number of UEs in the group g_u. Similarly, letting K_{g_d} be the number of UEs in the downlink group g_d, we can define the $g_{d,k} = k + \sum_{g_d'=1}^{g_d-1} K_{g_d'}$ to denote the kth downlink UE of the group g_d. Define $\mathbf{H}_{g_u} = \left[\mathbf{h}_{g_{u,1}},\cdots,\mathbf{h}_{g_{u,K_{g_u}}}\right]$ and $\mathbf{H}_{g_d} = \left[\mathbf{h}_{g_{d,1}},\cdots,\mathbf{h}_{g_{d,K_{g_d}}}\right]$ as the channel matrix from the uplink group g_u to the BS and that from the BS to the downlink group g_d, respectively, and define $\tilde{\mathbf{H}}_{g_u}$ and $\tilde{\mathbf{H}}_{g_d}$ as the corresponding beam-domain channel matrices.

During the data transmission phase, the uplink UEs transmit data to the BS, and meanwhile, the BS transmits data to the downlink UEs. Assuming the above UE grouping criteria, the received signals at the BS and downlink group g_d can be expressed as

$$\mathbf{y}_u = \sum_{g_u'\in\mathcal{G}_u} \mathbf{H}_{g_u'}\mathbf{s}_{g_u'} + \mathbf{H}_{SI}\sum_{g_d'\in\mathcal{G}_d} \mathbf{x}_{g_d'} + \mathbf{n}_u \tag{17}$$

$$\mathbf{y}_{g_d} = \mathbf{H}_{g_d}^H \sum_{g_d' \in G_d} \mathbf{x}_{g_d'} + \sum_{g_u' \in G_u} \mathbf{H}_{g_u' \to g_d} \mathbf{s}_{g_u'} + \mathbf{n}_{g_d} \tag{18}$$

where $\mathbf{n}_u$ and $\mathbf{n}_{g_d}$ denote the additive white Gaussian noises (AWGNs) with variance σ. $\mathbf{H}_{g_u' \to g_d}$ denotes the interference channel from uplink group g_u to downlink group g_d. Since the UEs have single antenna and are geographically distributed, the elements of $\mathbf{H}_{g_u' \to g_d}$ are assumed to be independent Gaussian random variables with zero mean. $\mathbf{s}_{g_u} \in \mathbb{C}^{K_{g_u} \times 1}$ denotes the transmit signal of uplink group g_u. $\mathbf{x}_{g_d} \in \mathbb{C}^{N \times 1}$ denotes the precoded transmit signal of the BS. In the BDFD scheme, the UEs of the downlink group g_d detect signal only on their active beam space. Therefore, we let $\mathbf{x}_{g_d} = \mathbf{F}^{\{:, B_{g_d}\}} \tilde{\mathbf{x}}_{g_d}$, where $\tilde{\mathbf{x}}_{g_d} \in \mathbb{C}^{b_d \times 1}$ is referred to as the beam-domain precoded transmit signal.

By multiplying both sides of (17) with $\left(\mathbf{F}^{\{:, B_{g_u}\}}\right)^H$ and using the definition $\mathbf{x}_{g_d} = \mathbf{F}^{\{:, B_{g_d}\}} \tilde{\mathbf{x}}_{g_d}$ on (18), we arrive at the beam-domain received signal at the BS from uplink UE group g_u and the beam-domain received signal at downlink group g_d

$$\tilde{\mathbf{y}}_{g_u} = \tilde{\mathbf{H}}_{g_u}^{\{B_{g_u}, :\}} \mathbf{s}_{g_u} + \sum_{g_u' \in G_u / \{g_u\}} \tilde{\mathbf{H}}_{g_u'}^{\{B_{g_u}, :\}} \mathbf{s}_{g_u'} + \sum_{g_d' \in G_d} \tilde{\mathbf{H}}_{SI}^{\left\{B_{g_u}, B_{g_d'}\right\}} \tilde{\mathbf{x}}_{g_d'} + \tilde{\mathbf{n}}_{g_u} \tag{19}$$

$$\tilde{\mathbf{y}}_{g_d} = \left(\tilde{\mathbf{H}}_{g_d}^{\{B_{g_d}, :\}}\right)^H \tilde{\mathbf{x}}_{g_d} + \sum_{g_d' \in G_d / \{g_d\}} \left(\tilde{\mathbf{H}}_{g_d}^{\left\{B_{g_d'}, :\right\}}\right)^H \tilde{\mathbf{x}}_{g_d'} + \sum_{g_u' \in G_u} \mathbf{H}_{g_u' \to g_d} \mathbf{s}_{g_u'} + \mathbf{n}_{g_d} \tag{20}$$

where $\tilde{\mathbf{n}}_{g_u} = \left(\mathbf{F}^{\{:, B_{g_u}\}}\right)^H \mathbf{n}_u$. Note that (20) is in fact the same with (18). To emphasize that (20) is the beam-domain received signal and be consistent with (19), we introduce the new notation $\tilde{\mathbf{y}}_{g_d}$. The second terms on the right-hand sides (RHSs) of (19) and (20) indicate the inter-group interferences (IGIs). The third term of RHS of (19) denotes the received SI. From (19), b_u should satisfy $b_u \geq K_{g_u}$ in order to support K_{g_u} independent data streams. Similarly, we require $b_d \geq K_{g_d}$ in downlink according to (20). As will be shown in Sect. 4.4, with simple UE scheduling, the interference from uplink UEs to downlink UEs can be made negligible compared to AWGN. Thus, we temporarily neglect this interference term in the analysis below. In this case, we build the optimality of the BDFD scheme using the following theorem.

Theorem 1 Assuming the UE grouping criteria is satisfied and the effective beam-domain channel matrices for all uplink and downlink groups are perfectly known, the BDFD scheme achieves the uplink and downlink sum capacities[5] simultaneously as the number of BS antennas approaches to infinity.

Proof Recalling the UE grouping Criterion 1 and Lemma 1, we can deduce that the IGI approaches to zero in the large N regime. Since we require $\left(\bigcap_{g_u \in G_u} B_{g_u}\right) \cap B_{SI,R} = \varnothing$ or $\left(\bigcap_{g_d \in G_d} B_{g_d}\right) \cap B_{SI,T} = \varnothing$ in UE grouping Criterion 2, the elements of $\tilde{\mathbf{H}}_{SI}^{\left\{B_{g_u}, B_{g_d'}\right\}}$ SI converge to zero[6] for very large N according to Lemma 2. Therefore, as $N \to \infty$ the beam-domain received signal $\tilde{\mathbf{y}}_{g_u}$ reduces to

$$\tilde{\mathbf{y}}_{g_u} = \tilde{\mathbf{H}}_{g_u}^{\{B_{g_u},:\}} \mathbf{s}_{g_u} + \tilde{\mathbf{n}}_{g_u} \tag{21}$$

Let $R_u^{\text{sum}}(P_u)$ be the uplink achievable sum rate with total power constraint P_u, and let $\Lambda_{g_u} = \mathbb{E}\left[\mathbf{s}_{g_u}\mathbf{s}_{g_u}^H\right]$ be the diagonal input covariance matrix of the uplink group g_u. Assuming the minimum mean square error with successive interference cancellation (MMSE-SIC) is employed to detect $\mathbf{s}_{g_u}$ from (23), the uplink achievable sum rate can be expressed as (22) [30] on the top of the next page, where the second step is based on the property $|\mathbf{I}+\mathbf{AB}| = |\mathbf{I}+\mathbf{BA}|$. The last step follows from Lemma 1, i.e., $\mathbf{H}_{g_u} = \mathbf{F}^{\{:,B_{g_u}\}}\tilde{\mathbf{H}}_{g_u}^{\{B_{g_u},:\}}$ is satisfied as $N \to \infty$. Note that the last line of (22) is exactly the uplink sum capacity [30].

$$\begin{aligned}
R_u^{\text{sum}}(P_u) &= \max_{\boldsymbol{\Lambda}_{g_u} \succcurlyeq 0,\ \sum\limits_{g_u \in G_u} \operatorname{tr}(\boldsymbol{\Lambda}_{g_u}) \le P_u} \sum_{g_u \in G_u} \log_2 \left| \mathbf{I}_{b_u} + \frac{1}{\sigma} \tilde{\mathbf{H}}_{g_u}^{\{B_{g_u}\}} \boldsymbol{\Lambda}_{g_u} \left(\tilde{\mathbf{H}}_{g_u}^{\{B_{g_u}\}}\right)^H \right| \\
&= \max_{\boldsymbol{\Lambda}_{g_u} \succcurlyeq 0,\ \sum\limits_{g_u \in G_u} \operatorname{tr}(\boldsymbol{\Lambda}_{g_u}) \le P_u} \sum_{g_u \in G_u} \log_2 \left| \mathbf{I}_N + \frac{1}{\sigma} \mathbf{F}^{\{B_{g_u}\}} \tilde{\mathbf{H}}_{g_u}^{\{B_{g_u}\}} \boldsymbol{\Lambda}_{g_u} \left(\tilde{\mathbf{H}}_{g_u}^{\{B_{g_u}\}}\right)^H \left(\mathbf{F}^{\{B_{g_u}\}}\right)^H \right| \\
&= \max_{\boldsymbol{\Lambda}_{g_u} \succcurlyeq 0,\ \sum\limits_{g_u \in G_u} \operatorname{tr}(\boldsymbol{\Lambda}_{g_u}) \le P_u} \sum_{g_u \in G_u} \log_2 \left| \mathbf{I}_N + \frac{1}{\sigma} \mathbf{H}_{g_u} \boldsymbol{\Lambda}_{g_u} \mathbf{H}_{g_u}^H \right|
\end{aligned} \tag{22}$$

Similarly, as $N \to \infty$, according to Criterion 1 and Lemma 1, the downlink beam-domain received signal $\tilde{\mathbf{y}}_{g_d}$ reduces to

$$\tilde{\mathbf{y}}_{g_d} = \left(\tilde{\mathbf{H}}_{g_d}^{\{B_{g_d},:\}}\right)^H \tilde{\mathbf{x}}_{g_d} + \mathbf{n}_{g_d} \tag{23}$$

[5]Herein, the uplink and downlink sum capacities indicate the maximum achievable rates of standard MIMO multiple access channel (MAC) and MIMO broadcasting (BC) channels, respectively [30].

[6]In (III-B) and (III-B), the terms of IGIs and SI are not exactly equal to zero for the general N. So we keep these terms in the equations.

Let $R_d^{\mathrm{sum}}(P_d)$ be the downlink achievable sum rate with total power constraint P_d. Assuming the beam-domain transmit signal $\tilde{\mathbf{x}}_{g_d}$ is generated according to the rule of dirty paper code and using the MAC-BC duality [30], we have

$$R_d^{\mathrm{sum}}(P_d) = \max_{\Lambda_{g_d} \succcurlyeq 0,\ \sum_{g_d \in G_d} \mathrm{tr}(\Lambda_{g_d}) \le P_d} \times \log_2 \left| \mathbf{I}_N + \frac{1}{\sigma} \sum_{g_d \in G_d} \mathbf{F}^{\{:, B_{g_d}\}} \tilde{\mathbf{H}}_{g_d}^{\{B_{g_d},:\}} \mathbf{\Lambda}_{g_d} \left(\tilde{\mathbf{H}}_{g_d}^{\{B_{g_d},:\}} \right)^H \left(\mathbf{F}^{\{:, B_{g_d}\}} \right)^H \right| \tag{24}$$

where $\mathbf{\Lambda}_{g_d}$ denotes the diagonal input covariance matrix of the dual MAC channel. To show the optimality of the BDFD scheme, we examine the following equality. Letting $G'_d \subset G_d$ and $G'_d \neq \varnothing$, for arbitrary $g_d \in G'_d$, we have

$$\begin{aligned}
&\left| \mathbf{I}_N + \tfrac{1}{\sigma} \sum_{g_d \in G_d} \mathbf{F}^{\{:, B_{g_d}\}} \tilde{\mathbf{H}}_{g_d}^{\{B_{g_d},:\}} \Lambda_{g_d} \left(\tilde{\mathbf{H}}_{g_d}^{\{B_{g_d},:\}} \right)^H \left(\mathbf{F}^{\{:, B_{g_d}\}} \right)^H \right| \\
&= \left| \mathbf{K}_{g_d} \right| \left| \mathbf{I}_N + \tfrac{1}{\sigma} \mathbf{K}_{g_d}^{-1} \sum_{g'_d \in G_d / \{g_d\}} \mathbf{F}^{\left\{:, B_{g'_d}\right\}} \tilde{\mathbf{H}}_{g'_d}^{\left\{B_{g'_d},:\right\}} \Lambda_{g'_d} \left(\tilde{\mathbf{H}}_{g'_d}^{\left\{B_{g'_d},:\right\}} \right)^H \left(\mathbf{F}^{\left\{:, B_{g'_d}\right\}} \right)^H \right| \\
&= \left| \mathbf{K}_{g_d} \right| \left| \mathbf{I}_N + \tfrac{1}{\sigma} \left(\mathbf{I}_N - \mathbf{F}^{\{:, B_{g_d}\}} \tilde{\mathbf{H}}_{g_d}^{\{B_{g_d},:\}} \left(\Lambda_{g_d}^{-1} + \left(\tilde{\mathbf{H}}_{g_d}^{\{B_{g_d},:\}} \right)^H \tilde{\mathbf{H}}_{g_d}^{\{B_{g_d},:\}} \right)^{-1} \left(\tilde{\mathbf{H}}_{g_d}^{\{B_{g_d},:\}} \right)^H \right.\right. \\
&\quad \left.\left. \times \left(\mathbf{F}^{\{:, B_{g_d}\}} \right)^H \right) \sum_{g'_d \in G_d / \{g_d\}} \mathbf{F}^{\left\{:, B_{g'_d}\right\}} \tilde{\mathbf{H}}_{g'_d}^{\left\{B_{g'_d},:\right\}} \Lambda_{g'_d} \left(\tilde{\mathbf{H}}_{g'_d}^{\left\{B_{g'_d},:\right\}} \right)^H \left(\mathbf{F}^{\left\{:, B_{g'_d}\right\}} \right)^H \right| \\
&= \left| \mathbf{K}_{g_d} \right| \left| \mathbf{I}_N + \tfrac{1}{\sigma} \sum_{g'_d \in G_d / \{g_d\}} \mathbf{F}^{\left\{:, B_{g'_d}\right\}} \tilde{\mathbf{H}}_{d,g'}^{\left\{B_{g'_d},:\right\}} \Lambda_{d,g'} \left(\tilde{\mathbf{H}}_{d,g'}^{\left\{B_{g'_d},:\right\}} \right)^H \left(\mathbf{F}^{\left\{:, B_{g'_d}\right\}} \right) \right|
\end{aligned} \tag{25}$$

where $\mathbf{K}_{g_d} = \mathbf{I}_N + \frac{1}{\sigma} \mathbf{F}^{\{:, B_{g_d}\}} \tilde{\mathbf{H}}_{g_d}^{\{B_{g_d},:\}} \Lambda_{g_d} \left(\tilde{\mathbf{H}}_{g_d}^{\{B_{g_d},:\}} \right)^H \left(\mathbf{F}^{\{B_{g_d},:\}} \right)^H$. The second step is obtained by applying the matrix inversion lemma on $\mathbf{K}_{g_d}^{-1}$ and the third step is based on Criterion 1. Using (25) repeatedly, we can rewrite the achievable downlink sum rate (24) as

$$\begin{aligned}
R_d^{\mathrm{sum}}(P_d) &= \max_{\Lambda_{g_d} \succcurlyeq 0,\ \sum_{g_d \in G_d} \mathrm{tr}(\Lambda_{g_d}) \le P_d} \\
&\quad \times \sum_{g_d \in G_d} \log_2 \left| \mathbf{I}_N + \tfrac{1}{\sigma} \mathbf{F}^{\{:, B_{g_d}\}} \tilde{\mathbf{H}}_{g_d}^{\{B_{g_d},:\}} \Lambda_{g_d} \left(\tilde{\mathbf{H}}_{g_d}^{\{B_{g_d},:\}} \right)^H \left(\mathbf{F}^{\{:, B_{g_d}\}} \right)^H \right| \\
&= \max_{\Lambda_{g_d} \succcurlyeq 0,\ \sum_{g_d \in G_d} \mathrm{tr}(\Lambda_{g_d}) \le P_d} \sum_{g_d \in G_d} \log_2 \left| \mathbf{I}_N + \tfrac{1}{\sigma} \mathbf{H}_{g_d} \Lambda_{g_d} \mathbf{H}_{g_d}^H \right|
\end{aligned} \tag{26}$$

which is exactly the sum capacity of the dual MAC channel with total power constraint P_d.

Theorem 1 reveals that only the reduced-dimension effective beam-domain CSI is enough for the BDFD scheme to achieve the uplink and downlink capacities simultaneously in the large N regime. Therefore, the BDFD scheme reduces the difficulty of channel acquisition. Note that conventional TDD/FDD massive MIMO can only achieve the uplink or downlink capacity on each time-frequency unit, even full CSI is available. Meanwhile, the BDFD scheme avoids the deployment of active SIC (which is hardware and energy costly) and does not need the instantaneous knowledge of the SI channel. However, in the practical application, the passive SIC may still be needed to suppress the residual SI under the SI channel-unaware environment.

Remark 1 Note that Theorem 1 is valid for single-cell system. In multicell system with FD BS, each transmission experiences more interferences compared to the single-cell situation, which include SI, UE-to-UE interference from both within the cell and neighboring cells, and BS-to-BS interference. To realize the gain of FD massive MIMO, efficient multicell interference mitigation technologies from different aspects, which may include UE scheduling, power control, and multiuser precoding with limited interference channel knowledge, should be studied.

4 Practical Implementation of BDFD Scheme

In this section, we consider several key components of BDFD scheme in the practical implementation, which include UE grouping, effective beam-domain channel acquisition, beam-domain data transmission, and interference control between uplink and downlink.

4.1 K-Means-Based UE Grouping

In real cellular system, UEs will not naturally partition in groups with exactly the same active beam set. In order to implement the BDFD scheme efficiently, the UEs with different active beam sets must be partitioned so that the UE grouping criteria are satisfied as close as possible. In this subsection, we propose a UE grouping scheme to achieve this task. Our scheme consists of the following three steps.

Step 1: Compute the active beam sets of all UEs and the SI channel based on the DOA/DOD information using the method in Sect. 3.1.

Step 2: The aim of the second step is to gather the UEs with similar active beam spaces into a group based on the K-means principle. Without loss of generality, we consider the uplink UEs and the operation for downlink UEs is similar. To apply the K-means algorithm, we need first to define the "distance" between UEs. In the proposed scheme, we employ the chordal distance between the active beam spaces of UEs. In particular, the distance between UEs k_u and k_u' can be expressed as

$$D_{\text{chordal}}\left(\mathbf{F}^{\{:,B_{k_u}\}}, \mathbf{F}^{\left\{:,B_{k_u'}\right\}}\right) = \left\| \mathbf{F}^{\{:,B_{k_u}\}}\left(\mathbf{F}^{\{:,B_{k_u}\}}\right)^H - \mathbf{F}^{\left\{:,B_{k_u'}\right\}}\left(\mathbf{F}^{\left\{:,B_{k_u'}\right\}}\right)^H \right\|^2$$
$$= \left|B_{k_u}\right| + \left|B_{k_u'}\right| - 2\left|B_{k_u} \cap B_{k_u'}\right| \qquad (27)$$

where the second equality is based on the orthogonality between columns of $\mathbf{F}^{\{:,B_{k_u}\}}$ and $\mathbf{F}^{\left\{:,B_{k_u'}\right\}}$. Moreover, for a group of UEs $\mathcal{U}$, the "centroid" of their active beam spaces is defined as [31]

$$\overline{\mathbf{F}} = \text{eig}_{b_u}\left\{\sum_{k_u \in U} \mathbf{F}^{\{:,B_{k_u}\}}\left(\mathbf{F}^{\{:,B_{k_u}\}}\right)^H\right\} \qquad (28)$$

where $\text{eig}_{b_u}\{\mathbf{A}\}$ indicates the unitary matrix whose columns are composed of b_u dominant eigenvectors of matrix $\mathbf{A}$. Since the columns of $\mathbf{F}$ are all eigenvectors of $\sum_{k_u \in \mathcal{U}} \mathbf{F}^{\{:,B_{k_u}\}}\left(\mathbf{F}^{\{:,B_{k_u}\}}\right)^H$, we can obtain $\overline{\mathbf{F}} = \mathbf{F}^{\{:,\overline{B}\}}$, where $\overline{B}$ denotes the active beam set of "centroid." By examining (28), it is easy to see that $\overline{B}$ can be expressed as

$$\overline{B} = \left\{f_1, f_2, \cdots, f_{b_d}\right\} \qquad (29)$$

where f_i is the index of the ith most frequent appeared beam in the sets $\left\{B_{k_u}\right\}_{k_u \in \mathcal{U}}$. With (27) and the notion of active beam set, we can conduct a very simple UE grouping algorithm based on the K-means principle, as shown in Algorithm 1. Note that as the output of Algorithm 1, the active beam set for group "centroid" is treated as the active beam set of that group.

Step 3: After step 2, we get a set of uplink UE groups, a set of downlink UE groups, and their active beam sets. To meet the SI cancellation condition in Criterion 2, for each uplink group, if its active beam set (denoted by B_u) is (partially) overlapped with $B_{SI,R}$, we update B_u as $B_u = B_u/(B_u \cap B_{SI,R})$. The active beam sets for downlink groups keep unchanged. Alternatively, we can also update the active beam set of downlink group (denoted by B_d) as $B_d = B_d/(B_d \cap B_{SI,T})$, if $B_d \cap B_{SI,T} \neq \varnothing$, while keeping the active beam sets for uplink groups unchanged. On the other hand, the active beam sets of different uplink/downlink UE groups after step 2 may also partially overlapped, which is not allowed according to Criterion 1. To deal with this problem, we further classify the uplink/downlink UE groups into several clusters so that the active beam sets of uplink/downlink UE groups in the same cluster are non-overlapping with certain guard interval. On certain time-frequency resource, only one uplink cluster and one downlink cluster are served using the BDFD scheme. Moreover, the UEs groups from different clusters are served using orthogonal time-frequency resources to eliminate the interference.

Remark 2 Here, we mention that orthogonal time-frequency resource allocation for different clusters does not mean more time-frequency resource consumption. The reason is that, in the cellular system, the number of UEs within each cell is commonly large. Thus it is impossible to serve all UEs using the same time-frequency resource. To access to the network, UEs which cannot be served simultaneously should be allocated to other time-frequency resources using technologies such as orthogonal frequency-division multiple access. Thus, the UE clustering operation can actually be viewed as an additional constraint on time-frequency resource allocation and will not degrade the system SE significantly.

Remark 3 When the active beam sets of uplink or downlink groups are (partially) overlapped with that of the SI channel, these groups get less beams due to Criterion 2. This may cause some problems in fairness between uplink and downlink UEs. Note that Criterion 2 can be satisfied by reducing the active beams of uplink UE groups or downlink groups in step 3. Therefore, if we reduce the active beams of uplink group to meet Criterion 2 in odd time slots, and reduce the active beams of downlink group to meet Criterion 2 in even time slots. The fairness can be improved to some extent. More intelligently, the resource allocation algorithm in time-frequency dimension can be investigated to achieve some kind of fairness (e.g., max-min fairness) among all UEs. This is interesting for future research.

4.2 Full-Duplex Effective Beam-Domain Channel Estimation

In this subsection, we propose a full-duplex channel estimation scheme to estimate the effective beam-domain channels. During the training phase, all the uplink UEs transmit pilot signals to the BS, and meanwhile, the BS transmits the pilot signals to the downlink UEs. Let $\Phi_u \in \mathbf{C}^{\tau_u \times \max\limits_{g_u \in G_u} K_{g_u}}$ be the orthogonal pilot sequence set for uplink training, where τ_u denotes the length of pilot sequence which satisfies $\tau_u \geq \max\limits_{g_u \in G_u} K_{g_u}$. The pilot sequences allocated for group g_u can be given by $\Phi_{g_u} = \Phi_u^{\{:,1:K_{g_u}\}}$. Meanwhile, let $\Phi_d \in \mathbf{C}^{\tau_d \times b_d}$ be the orthogonal downlink pilot sequences, where $\tau_d \geq b_d$ denotes the length of pilot sequence. The downlink pilot sequence for group g_d is precoded by multiplying the matrix $\mathbf{F}^{\{:,B_{g_d}\}}$. This operation is essential to suppress the IGI during training phase as will be seen below. The received pilot signals at the BS and the downlink group g_d can be expressed as

$$\mathbf{Y}_u = \sum_{g'_u \in G_u} \mathbf{H}_{g'_u} \Phi_{g'_u}^T + \mathbf{H}_{SI} \sum_{g'_d \in G_d} \mathbf{F}^{\left\{:,B_{g'_d}\right\}} \dot{\Phi}_d^T + \mathbf{N}_u \tag{30}$$

$$\mathbf{Y}_{g_d} = \mathbf{H}_{g_d}^H \sum_{g_d' \in G_d} \mathbf{F}^{\left\{:, B_{g_d'}\right\}} \Phi_d^T + \mathbf{N}_{g_d} \tag{31}$$

where $\dot{\Phi}_d = \Phi_d^{\{1:\tau_u\}}$ if $\tau_d \geq \tau_u$ and $\dot{\Phi}_d = \left[\Phi_d^T, 0_{(\tau_u - \tau_d) \times b_d}^T\right]^T$ if $\tau_d < \tau_u$. $\mathbf{N}_u$ and $\mathbf{N}_d$ denote the AWGNs with variance σ.

Uplink Effective Beam-Domain Channel Estimation

By multiplying both sides of (30) with $\left(\mathbf{F}^{\{:, B_{g_u}\}}\right)^H$, we arrive at the beam-domain receive pilot signal from uplink group g_u

$$\begin{aligned} \mathbf{Y}_{g_u} &= \left(\mathbf{F}^{\{:, B_{g_u}\}}\right)^H \mathbf{Y}_u \\ = \tilde{\mathbf{H}}_{g_u}^{\{B_{g_u},:\}} \Phi_{g_u}^T &+ \sum_{g_u' \in G_u / \{g_u\}} \tilde{\mathbf{H}}_{g_u'}^{\{B_{g_u},:\}} \Phi_{g_u'}^T + \sum_{g_d' \in G_d} \tilde{\mathbf{H}}_{SI}^{\left\{B_{g_u}, B_{g_d'}\right\}} \dot{\Phi}_d^T + \left(\mathbf{F}^{\{:, B_{g_u}\}}\right)^H \mathbf{N}_u \end{aligned} \tag{32}$$

With (32), the least squares (LS) estimator of the effective beam-domain channel vector for the uplink UE $g_{u,k}$ can be obtained as

$$\begin{aligned} \tilde{\mathbf{h}}_{g_{u,k},\mathrm{LS}}^{\{B_{g_u},:\}} &= \frac{1}{\tau_u p_u} \mathbf{Y}_{g_u} \Phi_u^* \mathbf{e}_k \\ = \tilde{\mathbf{h}}_{g_{u,k}}^{\{B_{g_u},:\}} + \sum_{g_u' \in G_u / \{g_u\}} \tilde{\mathbf{h}}_{g_{u,k}'}^{\{B_{g_u},:\}} &+ \frac{1}{\tau_u p_u} \sum_{g_d' \in G_d} \tilde{\mathbf{H}}_{SI}^{\left\{B_{g_u}, B_{g_d'}\right\}} \dot{\Phi}_d^T \Phi_u^* \mathbf{e}_k \\ &+ \frac{1}{\tau_u p_u} \left(\mathbf{F}^{\{:, B_{u,g}\}}\right)^H \mathbf{N}_u \Phi_u^* \mathbf{e}_k \end{aligned} \tag{33}$$

where p_u denotes the power of each uplink pilot symbol, i.e., $\left|\left[\Phi_{g_u}\right]_{i,j}\right|^2 = p_u$. The second term of RHS of (33) indicates the pilot contamination due to the use of same pilot sequences over all the uplink groups. The third term is the SI due to the simultaneous uplink and downlink training. Recalling Criterion 1 and Criterion 2 in the last section, and using Lemma 1 and Lemma 2, we can deduce that the pilot contamination and SI approach to zero in the large N regime.

In the practical scenario with finite number of BS antennas, the LS estimate in (33) can be further refined by a linear minimum mean square error (LMMSE) procedure to mitigate the residual pilot contamination and SI. Based on the general expression of LMMSE estimator [32, Ch. 12], the refined estimates can be expressed as

$$\begin{aligned}
\tilde{\mathbf{h}}_{g_{u,k},\mathrm{LM}}^{\{B_{g_u},:\}} &= \sum_{i=1}^{M_u} \int_{\theta_{g_{u,k,i}}^{\min}}^{\theta_{g_{u,k,i}}^{\max}} \left(\mathbf{F}^{\{:,B_{g_u}\}}\right)^H \mathbf{a}(\theta)\,\mathbf{a}^H(\theta)\,\mathbf{F}^{\{:,B_{g_u}\}} S_{g_{u,k,i}}(\theta)\,d\theta \\
&\times \left(\frac{\sigma}{\tau_u p_u}\mathbf{I}_{b_u} + \sum_{g_u'\in G_u}\sum_{i=1}^{M_u} \int_{\theta_{g_{u,k,i}}^{\min}}^{\theta_{g_{u,k,i}'}^{\max}} \left(\mathbf{F}^{\{:,B_{g_u}\}}\right)^H \mathbf{a}(\theta)\,\mathbf{a}^H(\theta)\,\mathbf{F}^{\{:,B_{g_u}\}} S_{g_{u,k}',i}(\theta)\,d\theta \right.\\
&+\left(\frac{1}{\tau_u p_u}\right)^2 \sum_{g_d',g_d''\in G_d}\sum_{i=1}^{M_{SI}} \int_{\theta_{R,i}^{\min}}^{\theta_{R,i}^{\max}}\int_{\theta_{T,i}^{\min}}^{\theta_{T,i}^{\max}} \mathbf{G}_{\theta_R,\theta_T}^{\left\{B_{g_u},B_{g_d'}\right\}} \Psi_k \left(\mathbf{G}_{\theta_R,\theta_T}^{\left\{B_{g_u},B_{g_d''}\right\}}\right)^H \\
&\left.\times\, S_{SI,i}(\theta_R,\theta_T)\,d\theta_R d\theta_T\right)^{-1} \tilde{\mathbf{h}}_{g_{u,k},\mathrm{LS}}^{\{B_{g_u},:\}}
\end{aligned} \tag{34}$$

where $\mathbf{G}_{\theta_R,\theta_T}^{\left\{B_{g_u},B_{g_d'}\right\}} \triangleq \left(\mathbf{F}^{\{:,B_{g_u}\}}\right)^H \mathbf{a}(\theta_R)\,\mathbf{a}^H(\theta_T)\,\mathbf{F}^{\left\{:,B_{g_d'}\right\}}$ and $\Psi_k \triangleq \left(\frac{1}{\tau_u p_u}\right)^2 \dot{\Phi}_d^T \Phi_u^* \mathbf{e}_k \mathbf{e}_k^H \Phi_u^T \dot{\Phi}_d^*$.

Downlink Effective Beam-Domain Channel Estimation

By exploiting the beam-domain presentation and using (31), the LS estimator for the effective beam-domain channel vector of the downlink UE $g_{d,k}$ can be obtained as

$$\begin{aligned}
\tilde{\mathbf{h}}_{g_{d,k},\mathrm{LS}}^{\{B_{g_d},:\}} &= \frac{1}{\tau_d p_d}\left(\mathbf{Y}_{g_d}\Phi_d^*\right)^H \mathbf{e}_k \\
&= \tilde{\mathbf{h}}_{g_{d,k}}^{\{B_{g_d},:\}} + \sum_{g_d'\in G_d} \tilde{\mathbf{h}}_{g_{d,k}}^{\left\{B_{g_d'},:\right\}} + \frac{1}{\tau_d p_d}\left(\mathbf{N}_{g_d}\Phi_d^*\right)^H \mathbf{e}_k
\end{aligned} \tag{35}$$

where p_d denotes the power of each downlink pilot symbol, i.e., $|[\Phi_d]_{i,j}|^2 = p_d$. The second term indicates the pilot contamination due to the use of same pilot sequences over all the downlink UE groups. Using Criterion 1 and Lemma 1, the pilot contamination converges to zero as $N \to \infty$. Similarly, we can refine the estimates with the LMMSE procedure, resulting in

$$\begin{aligned}
\tilde{\mathbf{h}}_{g_{d,k},\mathrm{LM}}^{\{B_{g_d},:\}} &= \sum_{g_d'\in G_d}\sum_{i=1}^{M_d} \int_{\theta_{g_{d,k,i}}^{\min}}^{\theta_{g_{d,k,i}}^{\max}} \left(\mathbf{F}^{\{:,B_{g_d}\}}\right)^H \mathbf{a}(\theta) \\
&\times \mathbf{a}^H(\theta)\,\mathbf{F}^{\left\{:,B_{g_d'}\right\}} S_{g_{d,k,i}}(\theta)\,d\theta \left(\sum_{g_d',g_d''\in G_d}\sum_{i=1}^{M_d} \int_{\theta_{g_{d,k,i}}^{\min}}^{\theta_{g_{d,k,i}}^{\max}} \left(\mathbf{F}^{\left\{:,B_{g_d'}\right\}}\right)^H \mathbf{a}(\theta)\right. \\
&\left.\times \mathbf{a}^H(\theta)\,\mathbf{F}^{\left\{:,B_{g_d''}\right\}} S_{g_{d,k,i}}(\theta)\,d\theta + \frac{\sigma}{\tau_d p_d}\right)^{-1} \tilde{\mathbf{h}}_{g_{d,k},\mathrm{LS}}^{\{B_{d,g},:\}}
\end{aligned} \tag{36}$$

Table 2 Minimum required lengths of pilot sequences in the proposed estimation scheme and conventional schemes

	BDFD	TDD massive MEMO (linear transceiver [3])	FDD massive MIMO (JSDM [8])	FD massive MIMO (linear transceiver [18])
Minimum length of pilot sequence	$\max\left\{\max_{g_u \in G_u} K_{g_u}, b_d\right\}$	$\sum_{g_u \in G_u} K_{g_u} + \sum_{g_d \in G_d} K_{g_d}$	$\max_{g_u \in G_u} K_{g_u} + b_d$ (Approximate)	$\sum_{g_u \in G_u} K_{g_u} + N$

To estimate the channels of all $\sum_{g_u \in G_u} K_{g_u}$ uplink UEs and $\sum_{g_d \in G_d} K_{g_d}$ downlink UEs, the minimum required lengths of pilot sequences in the proposed full-duplex estimation scheme and conventional schemes used in the TDD/FDD/FD massive MIMO systems are summarized in Table 2. It is seen that the proposed scheme improves the training efficiency significantly.

Example 2 Considering the channel model in Example 1, we have $b_d \approx 12$ if the BS is equipped with $N = 128$ antennas. If three uplink groups and three downlink groups are scheduled and each group contains five UEs, the minimum required length of pilot sequences in the proposed scheme is 12 (symbol times).

However, this number becomes 30, 17, and 143 (symbol times), respectively, in the reference schemes listed in Table 1. After downlink channel estimation, the estimated CSI should be feedback to BS in order to perform downlink transmission. This can affect the system from two aspects. First, the feedback error due to quantization error, noise, and feedback delay decreases the accuracy of downlink CSI. Moreover, CSI feedback increases the load of feedback channel and, hence, can degrade the overall system SE. However, the results in [33] showed that the CSI error (in term of mean-square error) due to imperfect feedback can be made much smaller than that caused by estimation error in downlink training phase, especially in the high signal-to-noise ratio (SNR) region. Moreover, since the effective downlink channel dimension is greatly reduced in proposed BDFD scheme, we assume that the additional load caused by CSI feedback is negligible when compared with the other feedback information. Therefore, for simplicity, we consider the optimistic situation of error-free CSI feedback and neglect the SE penalty due to feedback. A similar approach is also adopted in [8].

4.3 Beam-Domain Data Transmission and Achievable Rate with Noisy CSI

To keep the complexity low, we assume that the BS employs linear processing in the beam domain. In uplink, to detect the signals from group g_u, the BS combines the beam-domain received signal (Sect. 3.2) by multiplying the receive beamforming matrix $\mathbf{W}_{g_u} = \left[\mathbf{w}_{g_{u,1}}, \mathbf{w}_{g_{u,2}}, \cdots, \mathbf{w}_{g_{u,K_g}}\right] \in \mathbf{C}^{b_u \times K_g}$, i.e., $\dot{\mathbf{y}}_{g_u} = \mathbf{W}_{g_u}^H \tilde{\mathbf{y}}_{g_u}$. The kth entry of $\dot{\mathbf{y}}_{g_u}$

$$\begin{aligned}\dot{y}_{g_{u,k}} &= \mathbf{w}_{g_{u,k}}^H \tilde{\mathbf{h}}_{g_{u,k}}^{\{B_{g_u},:\}} s_{g_{u,k}} + \mathbf{w}_{u,gk}^H \sum_{k'=1,k'\neq k}^{K_{g_u}} \tilde{\mathbf{h}}_{g_{u,k'}}^{\{B_{g_u},:\}} s_{g_{u,k'}} \\ &+ \mathbf{w}_{g_{u,k}}^H \sum_{g'_u \in G_u/\{g_u\}} \tilde{\mathbf{H}}_{g'_u}^{\{B_{g_u},:\}} \mathbf{s}_{g'_u} + \mathbf{w}_{g_{u,k}}^H \sum_{g'_u \in G_d} \tilde{\mathbf{H}}_{SI}^{\left\{B_{g_u}, B_{g'_d}\right\}} \tilde{\mathbf{x}}_{g'_d} + \mathbf{w}_{g_{u,k}}^H \tilde{\mathbf{n}}_{g_u}\end{aligned} \tag{37}$$

is used to decode the symbol of UE $g_{u,k}$. In downlink, the intended signal of group g_d, i.e., $\mathbf{s}_{g_d} \in \mathbf{C}^{K_{g_d} \times 1}$, are precoded by the beamforming matrix $\mathbf{W}_{g_d} = \left[\mathbf{w}_{g_{d,1}}, \mathbf{w}_{g_{d,2}}, \cdots, \mathbf{w}_{g_{d,K_{g_d}}}\right] \in \mathbf{C}^{b_d \times K_{g_d}}$ in the beam domain. Thus, the beam-domain transmit signal vector for group g_d can be expressed as $\tilde{\mathbf{x}}_{g_d} = \mathbf{W}_{g_d} \mathbf{s}_{g_d}$. Using these on (III-B), the beam-domain received signal at UE $g_{d,k}$ can be expressed as

$$\begin{aligned}\tilde{y}_{g_{d,k}} &= \left(\tilde{\mathbf{h}}_{g_{d,k}}^{\{B_{g_d},:\}}\right)^H \mathbf{w}_{g_{d,k}} s_{g_{d,k}} + \left(\tilde{\mathbf{h}}_{g_{d,k}}^{\{B_{g_d},:\}}\right)^H \sum_{k'=1,k'\neq k}^{K} \mathbf{w}_{g_{d,k'}} s_{g_{d,k'}} \\ &+ \sum_{g'_d \in G_d/\{g_d\}} \left(\tilde{\mathbf{h}}_{g_{d,k}}^{\left\{B_{g'_d},:\right\}}\right)^H \mathbf{W}_{g'_d} \mathbf{s}_{g'_d} + n_{g_{d,k}}\end{aligned} \tag{38}$$

The optimal beamforming scheme to maximize the sum rate has been proved NP-hard [34]. Thus, we consider the suboptimal scheme to provide a bound on the system performance. In general, the (suboptimal) beamforming matrices can be designed with different criteria, e.g., maximizing the desired signal power which corresponds to the eigen beamforming or minimizing the inter-UE interference which corresponds to the zero-forcing (ZF) beamforming. In this work, we adopt the latter one since the ZF beamforming is known to approach the asymptotic limit of achievable rate faster as the number of BS antennas increases [2]. Assuming the channel estimators in (34) and (36), the transmit and receive beamforming matrices of the BS can be expressed as

$$\mathbf{W}_{g_u} = \tilde{\mathbf{H}}_{g_u,\mathrm{LM}}^{\{B_{g_u},:\}} \left(\left(\tilde{\mathbf{H}}_{g_u,\mathrm{LM}}^{\{B_{g_u},:\}}\right)^H \tilde{\mathbf{H}}_{g_u,\mathrm{LM}}^{\{B_{g_u},:\}}\right)^{-1}$$

$$\mathbf{W}_{g_d} = \tilde{\mathbf{H}}_{g_d,\mathrm{LM}}^{\{B_{g_d},:\}} \left(\left(\tilde{\mathbf{H}}_{g_d,\mathrm{LM}}^{\{B_{g_d},:\}}\right)^H \tilde{\mathbf{H}}_{g_d,\mathrm{LM}}^{\{B_{g_d},:\}}\right)^{-1} \Upsilon_{g_d}^{-1/2} \tag{39}$$

where Υ_{g_d} is a diagonal normalized matrix with $\left[\Upsilon_{g_d}\right]_{l,l} = \mathbf{e}_l^H\left(\left(\tilde{\mathbf{H}}_{g_d,\mathrm{LM}}^{\{B_{g_d},:\}}\right)^H \tilde{\mathbf{H}}_{g_d,\mathrm{LM}}^{\{B_{g_d},:\}}\right)^{-1}\mathbf{e}_l$. Due to the requirement of matrix inversion, the complexity of ZF beamforming becomes high when the number of UEs in each group is large. Some low complexity linear beamforming schemes, such as eigen beamforming, can be employed to deal with this problem at the cost of a few performance loss. To do so, we need just replace (39) with the beamforming matrices of these schemes. No other change is required.

According to (37) and (38) and using the bounding technique in [35], the average achievable rates at the uplink UE $g_{u,k}$ and downlink UE $g_{d,k}$ can be expressed as

$$R_{g_{u,k}}=\frac{T-\max\{\tau_u,\tau_d\}}{T}\log_2\left(1+\frac{p_{g_{u,k}}}{\mathbb{E}\left[\mathrm{CE}_{g_{u,k}}\right]+\mathbb{E}\left[\mathrm{IUI}_{g_{u,k}}\right]+\mathbb{E}\left[\mathrm{IGI}_{g_{u,k}}\right]+\mathbb{E}\left[\mathrm{SI}_{g_{u,k}}\right]+\mathbb{E}\left[\left\|\mathbf{w}_{u,g_k}\right\|^2\right]}\right)$$
$$R_{g_{d,k}} = \frac{T-\max\{\tau_u,\tau_d\}}{T}\log_2\left(1+\frac{p_{g_{d,k}}\mathbb{E}\left[\left[\Upsilon_{g_d}^{-1}\right]_{k,k}\right]}{\mathbb{E}\left[\mathrm{CE}_{g_{d,k}}\right]+\mathbb{E}\left[\mathrm{IUI}_{g_{d,k}}\right]+\mathbb{E}\left[\mathrm{IGI}_{g_{d,k}}\right]+1}\right) \tag{40}$$

where T denotes the channel coherent time. $p_{g_{u,k}} = \mathbb{E}\left[\left|s_{g_{u,k}}\right|^2\right]$ and $p_{g_{d,k}} = \mathbb{E}\left[\left|s_{g_{d,k}}\right|^2\right]$ denote the transmit powers. CEi, IUIi, IGIi, and SIi ($i \in \{g_{u,k}, g_{d,k}\}$) denote the powers of channel estimation error, inter-UE interference (IUI) within the group, IGI, and SI, respectively, whose expressions are summarized in Table 3.

Table 3 Expressions of powers of channel estimation error, IUI within the group, IGI, and SI

	Uplink	Downlink
Useful signal power	$p_{g_{u,k}}$	$p_{g_{d,k}}\left[\Upsilon_{g_d}^{-1}\right]_{k,k}$
Channel estimation error	$\mathrm{CE}_{g_{u,k}} = p_{g_{u,k}}\left\|\mathbf{w}_{g_{u,k}}^H\Delta\tilde{\mathbf{h}}_{g_{u,k}}^{\{B_{g_u},:\}}\right\|^2$	$\mathrm{CE}_{g_{d,k}} = p_{g_{d,k}}\left\|\left(\Delta\tilde{\mathbf{h}}_{g_{d,k},\mathrm{LM}}^{\{B_{g_d},:\}}\right)^H\mathbf{w}_{g_{d,k}}\right\|^2$
Inter-UE interference	$\mathrm{IUI}_{g_{u,k}} = \sum_{k'=1,k'\neq k}^{K_{g_u}} p_{g_{u,k'}}\left\|\mathbf{w}_{g_{u,k}}^H\Delta\tilde{\mathbf{h}}_{g_{u,k'},\mathrm{LM}}^{\{B_{g_u},:\}}\right\|^2$	$\mathrm{IUI}_{g_{d,k}} = \sum_{k'=1,k'\neq k}^{K_{g_d}} p_{g_{d,k'}}\left\|\left(\Delta\tilde{\mathbf{h}}_{g_{d,k},\mathrm{LM}}^{\{B_{g_d},:\}}\right)^H\mathbf{w}_{g_{d,k'}}\right\|^2$
Inter-group interference	$\mathrm{IGI}_{g_{u,k}} = \sum_{g'_u\in G_u/\{g_u\}}\left\Vert\mathbf{w}_{g_{u,k}}^H\tilde{\mathbf{H}}_{g'_u}^{\{B_{g_u},:\}}\Lambda_{g'_u}^{1/2}\right\Vert^2$	$\mathrm{IGI}_{g_{d,k}} = \sum_{g'_d\in G_d/\{g_d\}}\left\Vert\left(\tilde{\mathbf{h}}_{g_{d,k}}^{\{B_{g'_d},:\}}\right)^H\mathbf{W}_{g'_d}\Lambda_{g'_d}\right\Vert^2$
Self-interference	$\mathrm{SI}_{g_{u,k}} = \sum_{g'_d\in G_d}\left\Vert\mathbf{w}_{g_{u,k}}^H\tilde{\mathbf{H}}_{SI}^{\{B_{g_u},B_{g'_d}\}}\mathbf{W}_{g'_d}\Lambda_{g'_d}^{1/2}\right\Vert^2$	

The exact expressions of the average achievable rates are difficult to obtain under the considered channel model. Instead, in the following we focus on the question how do the above negative factors (i.e., channel estimation error, IUI, IGI, and SI) affect the achievable rate performance in the BDFD scheme. To answer this question, we present the scaling behaviors for powers of channel estimation error, IUI, IGI, and SI in the following theorem.

Theorem 2 Assume the cardinalities of the active beam sets scale linearly with N, i.e., $\lim_{N\to\infty} \frac{b_u}{N} > 0$ and $\lim_{N\to\infty} \frac{b_d}{N} > 0$. The scaling behaviors for average powers of channel estimation error, IUI, IGI, and SI in the large N regime are given by Table 4.

Proof See the Appendix.

Theorem 2 reveals that, in the BDFD scheme, the powers of IGI and SI decrease faster than other terms when the number of BS antennas increases. As a result, the effect of IGI and SI diminishes in the large N regime. In this sense, the BDFD scheme in fact decomposes the original system into several lower dimension uplink or downlink massive MIMO systems operating on the (asymptotically) orthogonal beam spaces. Another important observation from Theorem 2 is that the SI power decreases faster than $\mathcal{O}\left(N^{-1}\right)$ in the BDFD scheme. This is quite different from the FD massive MIMO with linear transceiver [18], where the SI power changes exactly with $\mathcal{O}\left(N^{-1}\right)$ in the large N regime. The reason is that, with the UE grouping criteria (Criterion 2), the signals of uplink or downlink groups occupy asymptotically orthogonal beam spaces with the SI. Thus, better SI suppression can be achieved in the BDFD scheme.

Remark 4 In theorem 2, we have assumed that b_u and b_d scale linearly with N. This is a standard assumption in the field of massive MIMO [8] in order to use the analytic tools developed for large-scale antenna systems. The assumption indicates that b_u and b_d, and hence the required length of pilot sequences, tend to infinity as $N \to \infty$, which is contrary to the purpose of this paper. However, in the practical implementation, the BS cannot be equipped with too many antennas due to the realistic constraints on hardware complexity and power consumption. With reasonable N, the training overhead is still low (See the Example 2 in Sect. 4.2).

Table 4 Scaling behaviors of channel estimation error, IUI within the group, IGI, and SI in the large N regime

	Useful signal power	Channel estimation error	Inter-UE interference	Inter-group interference	Self-interference
Uplink	$\mathcal{O}(1)$	$\mathcal{O}\left(N^{-1}\right)$	$\mathcal{O}\left(N^{-1}\right)$	$\begin{cases} \mathcal{O}\left(N^{\sigma_{\mathrm{IGI}}^{u}}\right) \\ \sigma_{\mathrm{IGI}}^{u} < -1 \end{cases}$	
Downlink	$\mathcal{O}(N)$	$\mathcal{O}(1)$	$\mathcal{O}(1)$	$\begin{cases} \mathcal{O}\left(N^{\sigma_{\mathrm{IGI}}^{d}}\right) \\ \sigma_{\mathrm{IGI}}^{d} < 0 \end{cases}$	$\begin{cases} \mathcal{O}\left(N^{\sigma_{\mathrm{SI}}}\right) \\ \sigma_{\mathrm{SI}} < -1 \end{cases}$

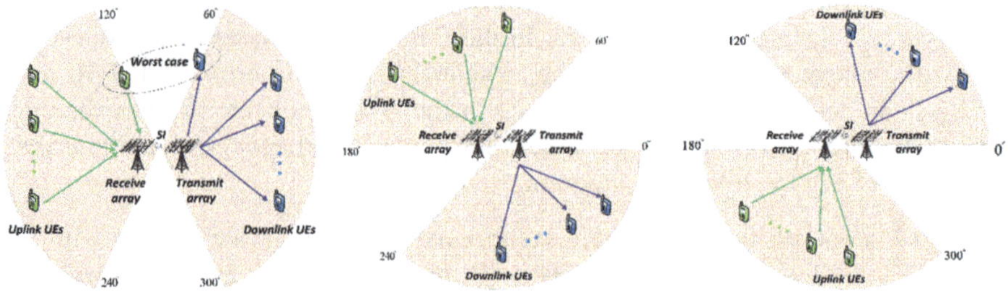

(a) (b) (c)

Fig. 4 Cell sectorization

4.4 Interference Control Between Uplink and Downlink

When multiple uplink UEs and downlink UEs are active simultaneously at the same frequency band, the resultant network suffers from increased interferences from uplink UEs to downlink UEs. One simple approach to alleviate the interference is cell sectorization. As shown in Fig. 4a, on the particular time-frequency resource, we only schedule the uplink UEs and downlink UEs in two opposite 120° sectors. This ensures that the uplink UE and downlink UE with small distance will not be scheduled on the same time-frequency resource. In the worst case where the uplink and downlink UEs are both located on the boundaries of the sectors as shown in Fig. 4a, the interference channel between uplink and downlink is still much weaker than the useful channel. For example, when the distance between BS and UEs is 300 m, the interference channel between two boundary UEs is 46 dB weaker than the useful channel according to the 3GPP LTE BS-to-UE and UE-to-UE path loss models [36, Table 6.4–1] (note that the UE-to-UE channel suffers from more path loss than the BS-to-UE channel even though the transmission distances are the same [36]). On the other hand, to cover the whole cell evenly, we can schedule the UEs in the rotated sectors, as shown in Fig. 4b, c, using different time-frequency resources.

5 Simulation Results

In this section, the performance of BDFD scheme is evaluated using the 3GPP LTE simulation model for macro-cell environment [36]. The simulation parameters are summarized in Table 5. It is assumed that the passive SIC scheme for infrastructure nodes proposed in [25] has been employed at the BS. In such scheme, the suppression is from two parts, namely, (i) the path loss introduced by the 20 m separation between transmit and receive antenna arrays and (ii) an additional cancellation of 45 dB provided by techniques, such as radio-frequency absorber material and cross-polarization. No other active SIC scheme is used.

Table 5 Simulation parameters

Parameter	Value
Bandwidth	20 MHz
Central frequency	2.4GHz
Thermal noise density	−174 dBm/Hz
Channel coherent time	200(Symbol times)
Path loss (BS-to-UE)	$2.7{+}42.8\log_{10}(R)[dB]$R: distance in meter
Path loss(UE-to-UE)	$55.78{+}40\log_{10}(R)[dB]$R: distance in meter
Number of scattering clusters	$M_u = M_d = 1$

We first consider a scenario where the uplink/downlink UEs gather perfectly in three groups and the DOA/DOD regions of UEs in each group are identical. We assume that each group contains five UEs. The DOA regions of three uplink groups are $[-33^{\circ}, -23^{\circ}]$, $[7^{\circ}, 17^{\circ}]$, and $[34^{\circ}, 44^{\circ}]$, respectively. Since we assume $Mu = 1$, the DOA region of uplink group is $[a, b]$ means $\left[\theta^{\min}_{g_{u,k},1}, \theta^{\max}_{g_{u,k},1}\right] = [a, b]$. Similarly, the DOD regions of three downlink groups are $[-39^{\circ}, -29^{\circ}]$, $[10^{\circ}, 20^{\circ}]$, and $[19^{\circ}, 30^{\circ}]$, respectively.

The DOA and DOD regions of SI channel are set to $[-15^{\circ}, -5^{\circ}]$, $[54^{\circ}, 66^{\circ}]$, and $[-25^{\circ}, -35^{\circ}]$, $[19^{\circ}, 30^{\circ}]$, respectively. The resulting active beam sets for all the groups satisfy the UE grouping criteria.

Figure 5 compares the SEs[7] of BDFD scheme, TDD massive MIMO with linear transceiver [3], FDD massive MIMO with JSDM [8], FD massive MIMO with linear transceiver [18], and FD massive MIMO with spatial SI suppression [37]. For the scheme with spatial SI suppression [37], the instantaneous CSI of SI channel is required at the BS in order to perform SI cancellation in spatial domain. With perfect effective beam-domain CSI, it is seen that the SE of BDFD scheme approaches the sum of uplink and downlink capacities as the number of BS antennas increases. With estimated effective beam-domain channels, the performance gap increases as N becomes larger. The reason is that, although the BDFD scheme can reduce the required length of pilot sequence significantly, the training overhead still increases linearly with N. Due to the same reason, when downlink reciprocity is available (at the cost of higher hardware complexity), the reference schemes in [18] and [37] achieve better SE over the BDFD scheme in the large N region. On the other hand, significant SE gain can be achieved by the BDFD scheme over the TDD and FDD massive MIMO systems. Interestingly, the performance gain can even be greater than $2\times$ (e.g., $2.08\times$ gain is observed over the TDD massive MIMO when $N = 200$), which is impossible in the conventional FD system. This is because the TDD massive MIMO spends more resource for pilot signaling as discussed in Sect.

[7]The SEs of BDFD scheme and FD massive MIMO with linear transceiver are defined as the sum of achievable rates of all uplink and downlink UEs. The SEs of TDD/FDD massive MIMO are defined in the same way but penalized by a factor of 1/2 due to the orthogonal uplink/downlink resource allocation.

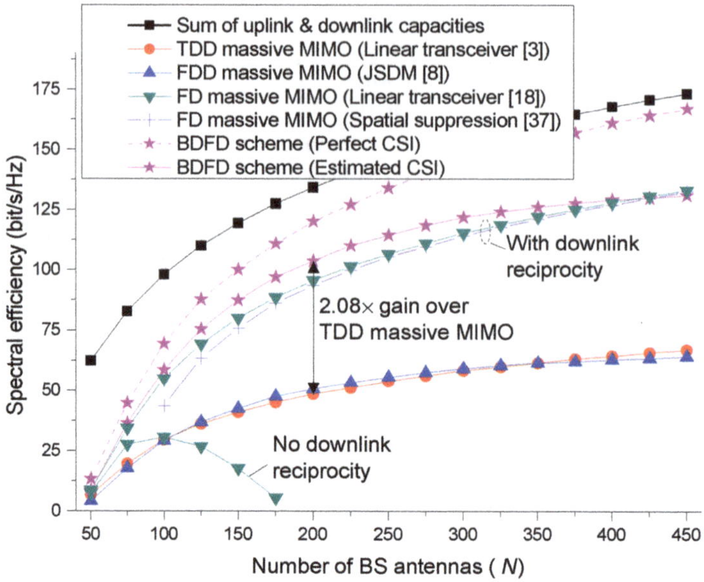

Fig. 5 Spectral efficiency with perfect UE grouping. The transmit powers of BS and uplink UEs are 20.2 dBm. The distance between UEs and BS is set to 500 m. This setup ensures the average uplink/downlink receive SNR is 3 dB

4.2. At last, without downlink reciprocity, it is observed that the FD massive MIMO with linear transceiver becomes infeasible if the number of BS antennas exceeds 175, since almost all the time resource is allocated for downlink training.

To examine the scaling results in Theorem 2, Fig. 6 simulates the average powers of useful signal, channel estimation error, IUI, IGI, and SI in the BDFD scheme. From the figure, it is seen that the powers of IGI and SI decrease faster than other negative factors in the large N regime, which coincides with Theorem 2. Then, in Figs. 7 and 8, we consider a more realistic scenario where the UEs are not naturally partitioned in groups with exactly the same active beam set. We assume that 50 uplink UEs and 50 downlink UEs are located in two opposite 120° sectors, as shown in Fig. 4a. The BS is equipped with $N = 128$ transmit/receive antennas. The signal of each uplink/downlink UE is within a 10° DOA/DOD region which is randomly distributed in the sectors. The distance between uplink/downlink UE and BS is randomly distributed in the interval [200, 1000] m. After UE grouping, three UE clusters are formed using the method in Sect. 4.1, and the UEs groups in different clusters are served with orthogonal time-frequency resources. Without loss of generality, the active beam sets of uplink groups which are (partially) overlapped that of the SI channel are updated using the method in Sect. 4.1 (step 3).

Figure 7 depicts the SE of BDFD scheme as a function of average receive SNR. Since the UE groups are divided into three clusters, the SE is defined as the average of SEs for three clusters. The number of scattering cluster for SI channel is set to

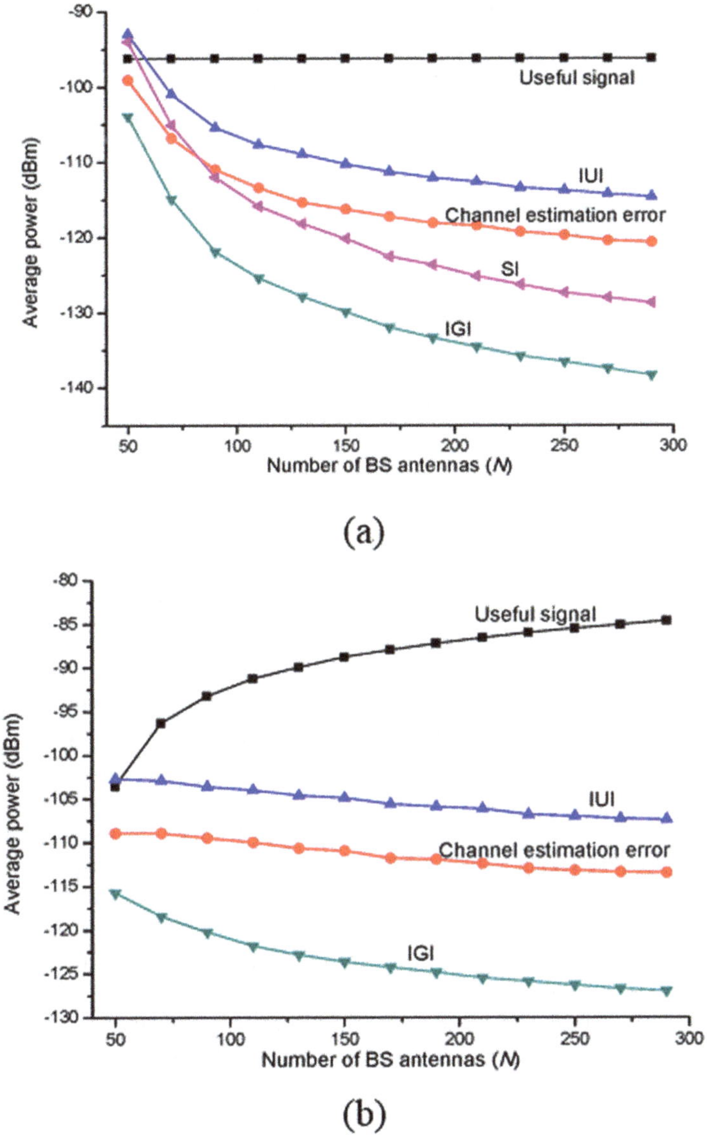

Fig. 6 Average powers of useful signal, channel estimation error, and interferences in the BDFD scheme. The simulation setup is the same with Fig. 5. (**a**) Uplink. (**b**) Downlink

$M_{SI} = 2$. The DOA and DOD regions of SI channel are $[-15^\circ, -5^\circ]$, $[54^\circ, 66^\circ]$, and $[-25^\circ, -35^\circ]$, $[19^\circ, 30^\circ]$, respectively. Again, it is seen that the BDFD scheme achieves the best performance. In particular, the BDFD scheme achieves 1.80$\times$ and 1.87 $\times$ SE gain over the TDD massive MIMO when the SNRs are 3 dB and 12 dB,

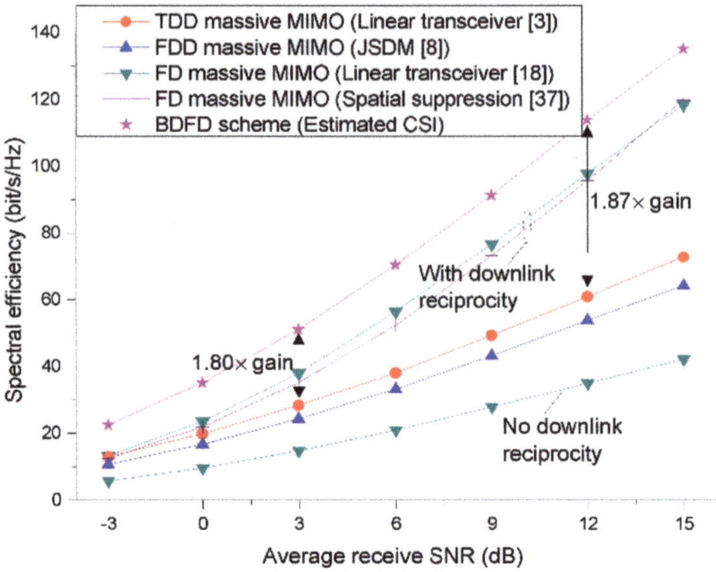

Fig. 7 Spectral efficiency of BDFD scheme with imperfect UE grouping for different average receive SNRs

respectively. The performance gain is generally smaller compared with that in Fig. 5. The reason is that, different from Fig. 5, the UEs in each group may not have exactly the same active beam set. Thus, not all the beams in the active beam set of that group can be fully used by all the UEs. This will result in some performance degradation. Moreover, there is no performance floor in the large SNR region for the BDFD scheme, even the interference from uplink UE to downlink exist. At last, we mention that, with the UE scheduling scheme in Sect. 4.4, the powers of interferences from uplink UEs to downlink UEs are much smaller than the background noise. That is why no obvious performance floor is observed for BDFD scheme in the large SNR region.

In the previous simulations, the number of scattering clusters for SI channel is fixed to $M_{SI} = 2$. In Fig. 8, we consider the SE of BDFD scheme with larger M_{SI}. In particular, we let M_{SI} increase from 2 to 10. The SI signal from each scattering cluster is within a 10° DOA/DOD region which is randomly distributed $[-90^{\circ}, 90^{\circ}]$. It is seen the SE of BDFD scheme approaches to that of the TDD/FDD massive MIMO as M_{SI} increases. This is because the numbers of uplink groups and the active beams for each uplink group decrease according to algorithm in Sect. 4.1. In fact, for large M_{SI}, the performance gain of BDFD scheme over TDD/FDD massive MIMO is mainly due to the saving in the training resources. Moreover, since the scheme in [37] cancels the SI completely using the instantaneous CSI, it achieves better SE when $M_{SI} \geq 6$. However, the gain can be realized only when the instantaneous SI channel can be efficiently estimated.

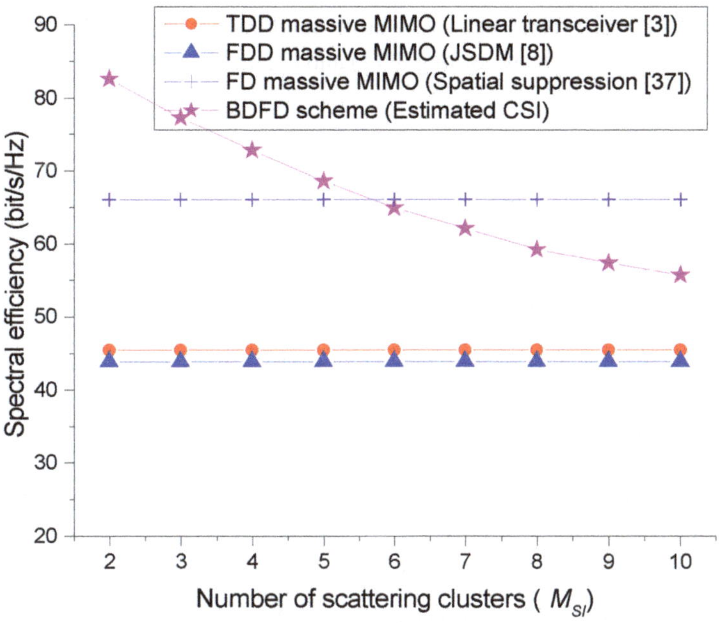

Fig. 8 Spectral efficiency of BDFD scheme for different numbers of scattering clusters for SI channel

6 Conclusion

This paper proposes a BDFD massive MIMO scheme to realize CCUD transmission in the cellular system. By exploiting the compressibility of beam-domain channel, the BDFD scheme can eliminate SI due to CCUD transmission efficiently. The simulation results show that the BDFD massive MIMO scheme outperforms the TDD/FDD massive MIMO and FD massive MIMO with linear transceiver significantly in the macro-cell environment. Due to the above advantages, we suggest BDFD massive MIMO as a potential enabling technology for evolution toward future wireless cellular system.

Acknowledgment This work was supported in part by National Natural Science Foundation of China under Grant 61671472 and in part by Jiangsu Province Natural Science Foundation under Grant BK20160079.

Appendix

As in [6], we assume that the number of channel paths is very large within the DOA/DOD regions. As a result, we can assume that the uplink/downlink channel is Gaussian distributed from the law of large numbers, i.e., $\mathbf{h}_{g_{u,k}} \sim \mathcal{CN}\left(0, \mathbf{C}_{g_{u,k}}\right)$ and $\mathbf{h}_{g_{d,k}} \sim \mathcal{CN}\left(0, \mathbf{C}_{g_{d,k}}\right)$. According to (1) and (2), the correlation matrices can be expressed as

$$\mathbf{C}_{g_{u,k}} = \mathbb{E}\left[\mathbf{h}_{g_{u,k}}\mathbf{h}_{g_{u,k}}^H\right] = \sum_{i=1}^{M_u}\int_{\theta_{g_{u,k},i}^{\min}}^{\theta_{g_{u,k},i}^{\max}} \mathbf{a}(\theta)\,\mathbf{a}^H(\theta)\, S_{g_{u,k},i}(\theta)\, d\theta$$

$$\mathbf{C}_{g_{d,k}} = \mathbb{E}\left[\mathbf{h}_{g_{d,k}}\mathbf{h}_{g_{d,k}}^H\right] = \sum_{i=1}^{M_d}\int_{\theta_{g_{d,k},i}^{\min}}^{\theta_{g_{d,k},i}^{\max}} \mathbf{a}(\theta)\,\mathbf{a}^H(\theta)\, S_{g_{d,k},i}(\theta)\, d\theta \tag{41}$$

Based on the above assumption, we have $\tilde{\mathbf{h}}_{g_{u,k}}^{\left\{B_{g'_u},:\right\}} \sim \mathcal{CN}\left(0, \tilde{\mathbf{C}}_{g_{u,k}}^{g'_u}\right)$ and $\tilde{\mathbf{h}}_{g_{d,k}}^{\left\{B_{g'_d},:\right\}} \sim \mathcal{CN}\left(0, \tilde{\mathbf{C}}_{g_{d,k}}^{g'_d}\right)$, where $\tilde{\mathbf{C}}_{g_{u,k}}^{g'_u} = \left(\mathbf{F}^{\left\{:,B_{g'_u}\right\}}\right)^H \mathbf{C}_{g_{u,k}}\mathbf{F}^{\left\{:,B_{g'_u}\right\}}$ and $\tilde{\mathbf{C}}_{g_{d,k}}^{g'_d} = \left(\mathbf{F}^{\left\{:,B_{g'_d}\right\}}\right)^H \mathbf{C}_{g_{d,k}}\mathbf{F}^{\left\{:,B_{g'_d}\right\}}$. With this and the standard result for LMMSE estimator [32, Ch. 12], the LMMSE estimates for uplink and downlink channels have distributions $\tilde{\mathbf{h}}_{g_{u,k},\mathrm{LM}}^{\left\{B_{g_u},:\right\}} \sim \mathcal{CN}\left(0, \tilde{\mathbf{C}}_{g_{u,k},\mathrm{LM}}^{g_u}\right)$ and $\tilde{\mathbf{h}}_{g_{d,k},\mathrm{LM}}^{\left\{B_{g_d},:\right\}} \sim CN\left(0, \tilde{\mathbf{C}}_{g_{d,k},\mathrm{LM}}^{g_d}\right)$, where the correlation matrices can be expressed as

$$\begin{aligned}\tilde{\mathbf{C}}_{g_{u,k},\mathrm{LM}}^{g_u} = \tilde{\mathbf{C}}_{g_{u,k}}^{g_u}\Bigg(&\frac{\sigma}{\tau_u p_u}\mathbf{I}_{b_u} + \sum_{g'_u\in G_u}\tilde{\mathbf{C}}_{g'_{u,k}}^{g_u} + \sum_{g'_d,g''_d\in G_d}\sum_{i=1}^{M_{SI}}\int_{\theta_{R,i}^{\min}}^{\theta_{R,i}^{\max}}\int_{\theta_{T,i}^{\min}}^{\theta_{T,i}^{\max}}\mathbf{G}_{\theta_R,\theta_T}^{\left\{B_{u,g},B_{d,g'}\right\}} \\ &\times\Psi_k\left(\left(\mathbf{G}_{\theta_R,\theta_T}^{\left\{B_{u,g},B_{d,g''}\right\}}\right)^H S_{SI,i}(\theta_R,\theta_T)\,d\theta_R d\theta_T\right)\Bigg)^{-1}\tilde{\mathbf{C}}_{g_{u,k}}^{g_u}\end{aligned}$$

$$\begin{aligned}\tilde{\mathbf{C}}_{g_{d,k},\mathrm{LM}}^{g_d} = &\sum_{g'_d\in G_d}\sum_{i=1}^{M_d}\int_{\theta_{g_{d,k},i}^{\min}}^{\theta_{g_{d,k},i}^{\max}}\left(\mathbf{F}^{\{:,B_{g_d}\}}\right)^H\mathbf{a}(\theta)\,\mathbf{a}^H(\theta)\,\mathbf{F}^{\left\{:,B_{g'_d}\right\}}S_{g_{d,k},i}(\theta)\,d\theta \\ \times&\left(\sum_{g',g''\in G_d}\sum_{i=1}^{M_d}\int_{\theta_{g_{d,k},i}^{\min}}^{\theta_{g_{d,k},i}^{\max}}\left(\mathbf{F}^{\left\{:,B_{g'_d}\right\}}\right)^H\mathbf{a}(\theta)\,\mathbf{a}^H(\theta)\,\mathbf{F}^{\left\{:,B_{g''_d}\right\}}S_{g_{d,k},i}(\theta)\,d\theta+\frac{\sigma}{\tau_d p_d}\mathbf{I}_{b_d}\right)^{-1} \\ \times&\sum_{g'_d\in G_d}\sum_{i=1}^{M_d}\int_{\theta_{g_{d,k},i}^{\min}}^{\theta_{g_{d,k},i}^{\max}}\left(\mathbf{F}^{\left\{:,B_{g'_d}\right\}}\right)^H\mathbf{a}(\theta)\,\mathbf{a}^H(\theta)\,\mathbf{F}^{\{:,B_{g_d}\}}S_{g_{d,k},i}(\theta)\,d\theta\end{aligned} \tag{42}$$

Moreover, as $N \to \infty$, we have [3]

$$\frac{1}{N}\left(\tilde{\mathbf{h}}_{g_{u,k},\mathrm{LM}}^{\{B_{g_u},:\}}\right)^H \tilde{\mathbf{h}}_{g_{u,k},\mathrm{LM}}^{\{B_{g_u},:\}} = \frac{1}{N}\mathrm{tr}\left(\tilde{\mathbf{C}}_{g_{u,k},\mathrm{LM}}^{g_u}\right)$$

$$\frac{1}{N}\left(\tilde{\mathbf{h}}_{g_{d,k},\mathrm{LM}}^{\{B_{g_d},:\}}\right)^H \tilde{\mathbf{h}}_{g_{d,k},\mathrm{LM}}^{\{B_{g_d},:\}} = \frac{1}{N}\mathrm{tr}\left(\tilde{\mathbf{C}}_{g_{d,k},\mathrm{LM}}^{g_d}\right) \tag{43}$$

With (41), (42), and (43), the scaling behaviors for average powers of useful signal, channel estimation, IUI, and IGI can be readily obtained by using the technique in [38, Proof of Theorem 4] as shown in (44) at the bottom of the page. Then we focus on the scaling behavior of SI power. According to Table 3, the asymptotic result in (43) and the property tr(**AB**) = tr(**BA**), we can rewrite the average SI power as

$$\mathbb{E}\left[\mathrm{SI}_{g_{u,k}}\right] = \left(\mathrm{tr}\left(\tilde{\mathbf{C}}_{g_{u,k},\mathrm{LM}}^{g_u}\right)\right)^{-2} \sum_{g'_d \in G_d} \sum_{k'=1}^{K_{g_d}} p_{g'_{d,k'}} \left(\mathrm{tr}\left(\tilde{\mathbf{C}}_{g'_{d,k'},\mathrm{LM}}^{g'_d}\right)\right)^{-1}$$
$$\times \underbrace{\mathrm{tr}\left(\mathbb{E}\left[\tilde{\mathbf{h}}_{g_{u,k},\mathrm{LM}}^{\{B_{g_u},:\}}\left(\tilde{\mathbf{h}}_{g_{u,k},\mathrm{LM}}^{\{B_{g_u},:\}}\right)^H \tilde{\mathbf{H}}_{SI}^{\left\{B_{g_u},B_{g'_d}\right\}} \tilde{\mathbf{h}}_{g'_{d,k'},\mathrm{LM}}^{\left\{B_{g'_d},:\right\}}\left(\tilde{\mathbf{h}}_{g'_{d,k'},\mathrm{LM}}^{\left\{B_{g'_d},:\right\}}\right)^H \left(\tilde{\mathbf{H}}_{SI}^{\left\{B_{g_u},B_{g'_d}\right\}}\right)^H\right]\right)}_{X_{g_{u,k},g'_{d,k}}} \tag{44}$$

The expression of $X_{g_{u,k},g'_{d,k}}$ can be rewritten as

$$\begin{aligned}
X_{g_{u,k},g'_{d,k}} &= \mathrm{tr}\left(\mathbb{E}\left[\tilde{\mathbf{C}}_{g_{u,k},\mathrm{LM}}^{g_u}\tilde{\mathbf{H}}_{SI}^{\left\{B_{g_u},B_{g'_d}\right\}}\tilde{\mathbf{C}}_{g_{d,k'},\mathrm{LM}}^{g_d}\left(\tilde{\mathbf{H}}_{SI}^{\left\{B_{g_u},B_{g'_d}\right\}}\right)^H\right]\right) \\
&\leq \mathrm{tr}\left(\mathbb{E}\left[\tilde{\mathbf{C}}_{g_{u,k}}^{g_u}\tilde{\mathbf{H}}_{SI}^{\left\{B_{g_u},B_{g'_d}\right\}}\tilde{\mathbf{C}}_{g_{d,k'}}^{g_d}\left(\tilde{\mathbf{H}}_{SI}^{\left\{B_{g_u},B_{g'_d}\right\}}\right)^H\right]\right) \\
&= \mathrm{tr}\left(\tilde{\mathbf{C}}_{g_{u,k}}^{g_u}\sum_{i=1}^{M_{SI}}\int_{\theta_{R,i}^{\min}}^{\theta_{R,i}^{\max}}\int_{\theta_{T,i}^{\min}}^{\theta_{T,i}^{\max}} S_{SI,i}\left(\theta_R,\theta_T\right)\left(\mathbf{F}^{\{:,B_{g_u}\}}\right)^H \mathbf{a}\left(\theta_R\right)\mathbf{a}^H\left(\theta_T\right)\right. \\
&\quad \left.\times \mathbf{F}^{\{:,B_{d,g}\}}\tilde{\mathbf{C}}_{g'_{d,k'}}^{g'_d}\left(\mathbf{F}^{\left\{:,B_{g'_d}\right\}}\right)^H \mathbf{a}\left(\theta_T\right)\mathbf{a}^H\left(\theta_R\right)\mathbf{F}^{\{:,B_{g_u}\}}d\theta_R d\theta_T\right) \\
&\leq I_R \times I_T \max_{\theta_R \in \bigcup_{i=1}^{M_{SI}}\left[\theta_{R,i}^{\min},\theta_{R,i}^{\max}\right],\theta_T \in \bigcup_{i=1}^{M_{SI}}\left[\theta_{T,i}^{\min},\theta_{T,i}^{\max}\right]} S_{SI,i}\left(\theta_R,\theta_T\right)
\end{aligned} \tag{45}$$

where

$$I_R = \int\limits_{\theta_R \in \bigcup_{i=1}^{M_{SI}} \left[\theta_{R,i}^{\min}, \theta_{R,i}^{\max}\right]} \mathbf{a}^H(\theta_R)\, \mathbf{F}^{\{:,B_{gu}\}} \tilde{\mathbf{C}}_{g_{u,k}}^{g_u} \left(\mathbf{F}^{\{:,B_{gu}\}}\right)^H \mathbf{a}(\theta_R)\, d\theta_R$$

$$I_T = \int\limits_{\theta_T \in \bigcup_{i=1}^{M_{SI}} \left[\theta_{T,i}^{\min}, \theta_{T,i}^{\max}\right]} \mathbf{a}^H(\theta_T)\, \mathbf{F}^{\left\{:,B_{g_d'}\right\}} \tilde{\mathbf{C}}_{g_{d',k'}}^{g_d'} \left(\mathbf{F}^{\left\{:,B_{g_d'}\right\}}\right)^H \mathbf{a}(\theta_T)\, d\theta_T \tag{46}$$

In (45), the first step is based on the independence between $\tilde{\mathbf{h}}_{g_{u,k},\mathrm{LM}}^{\{B_{gu},:\}}$, $\tilde{\mathbf{h}}_{g_{d,k}',\mathrm{LM}}^{\left\{B_{g_d'},:\right\}}$, and $\tilde{\mathbf{H}}_{SI}^{\left\{B_{gu},B_{g_d'}\right\}}$. The second step is based on the relation $\tilde{\mathbf{C}}_{g_{m,k},\mathrm{LM}}^{g_m} = \tilde{\mathbf{C}}_{g_{m,k}}^{g_m} - \left(\tilde{\mathbf{C}}_{g_{m,k}}^{g_m} - \tilde{\mathbf{C}}_{g_{m,k},\mathrm{LM}}^{g_m}\right)(m \in \{u,d\})$ and the positive definiteness of $\tilde{\mathbf{C}}_{g_{m,k}}^{g_m}$ and $\tilde{\mathbf{C}}_{g_{m,k}}^{g_m} - \tilde{\mathbf{C}}_{g_{m,k},\mathrm{LM}}^{g_m}$. The third step is obtained by using the equation $\tilde{\mathbf{H}}_{SI}^{\left\{B_{gu},B_{g_d'}\right\}} = \left(\mathbf{F}^{\{:,B_{gu}\}}\right)^H \mathbf{H}_{SI}\mathbf{F}^{\left\{:,B_{g_d'}\right\}}$ and the SI channel model (3).

By substituting (41) into (46), the integral I_R can be rewritten as

$$\begin{aligned}
\frac{1}{N} I_R &= \frac{1}{N} \sum_{i=1}^{M_u} \int_{\theta_{g_{u,k},i}^{\min}}^{\theta_{g_{u,k},i}^{\max}} S_{g_{u,k},i}(\theta) \int\limits_{\theta_R \in \bigcup_{i=1}^{M_{SI}} \left[\theta_{R,i}^{\min}, \theta_{R,i}^{\max}\right]} \mathbf{a}^H(\theta_R)\, \mathbf{a}(\theta)\, \mathbf{a}^H(\theta)\, \mathbf{a}(\theta_R)\, d\theta_R d\theta \\
&= N \sum_{i=1}^{M_u} \int_{\theta_{g_{u,k},i}^{\min}}^{\theta_{g_{u,k},i}^{\max}} S_{g_{u,k},i}(\theta) \int\limits_{\theta_R \in \bigcup_{i=1}^{M_{SI}} \left[\theta_{R,i}^{\min}, \theta_{R,i}^{\max}\right]} \mathrm{asinc}_N^2\left(\tfrac{d}{\lambda}\sin\theta_R - \tfrac{d}{\lambda}\sin\theta\right) d\theta_R d\theta
\end{aligned} \tag{47}$$

Note that according to Lemma 1 and Lemma 2, we have $\cap_{i=1}^{M_u}\left[\theta_{g_{u,k},i}^{\min}, \theta_{g_{u,k},i}^{\max}\right] \cap \cap_{i=1}^{M_{SI}}\left[\theta_{R,i}^{\min}, \theta_{R,i}^{\max}\right] = \varnothing$ or $\cap_{i=1}^{M_d}\left[\theta_{g_{d,k},i}^{\min}, \theta_{g_{d,k},i}^{\max}\right] \cap \cap_{i=1}^{M_{SI}}\left[\theta_{T,i}^{\min}, \theta_{T,i}^{\max}\right] = \varnothing$, otherwise, Criterion 2 will be violated. If $\cap_{i=1}^{M_u}\left[\theta_{g_{u,k},i}^{\min}, \theta_{g_{u,k},i}^{\max}\right] \cap \cap_{i=1}^{M_{SI}}\left[\theta_{R,i}^{\min}, \theta_{R,i}^{\max}\right] = \varnothing$, with a same procedure as that in the proof of Lemma 1, we can obtain $\frac{1}{N} I_R < \mathcal{O}(1)$ or $I_R < \mathcal{O}(N)$ as $N \to \infty$; otherwise, $I_R = \mathcal{O}(N)$. In the same way, we can prove that $I_T < \mathcal{O}(N)$ if $\cap_{i=1}^{M_d}\left[\theta_{g_{d,k},i}^{\min}, \theta_{g_{d,k},i}^{\max}\right] \cap \cap_{i=1}^{M_{SI}}\left[\theta_{T,i}^{\min}, \theta_{T,i}^{\max}\right] = \varnothing$, and $I_T = \mathcal{O}(N)$ otherwise. Combining the results in the above, we have $I_R \times I_T < \mathcal{O}\left(N^2\right)$.

Moreover, it has been shown in [38] that $\mathrm{tr}\left(\tilde{\mathbf{C}}^{g_u}_{g_{u,k},\mathrm{LM}}\right)$ and $\mathrm{tr}\left(\tilde{\mathbf{C}}^{g'_d}_{g_{d'},k',\mathrm{LM}}\right)$ scale with $\mathcal{O}(N)$ as $N \to \infty$. Using these results on (44), we have $\mathbb{E}\left[\mathrm{SI}_{g_{u,k}}\right] < \mathcal{O}\left(N^{-1}\right)$, which is exactly the result in Table 3.

References

1. T.L. Marzetta, Noncooperative cellular wireless with unlimited numbers of BS antennas. IEEE Trans. Wirel. Commun. **9**(11), 3590–3600 (2010)
2. H.Q. Ngo, E.G. Larsson, T.L. Marzetta, Energy and spectral efficiency of very large multiuser MIMO systems. IEEE Trans. Commun. **61**(4), 1436–1449 (2013)
3. J. Hoydis, S. Brink, M. Debbah, Massive MIMO in UL/DL of cellular networks: how many antennas do we need? IEEE J. Sel. Areas Commun. **31**(2), 160–171 (2013)
4. J. Zhang, C.K. Wen, S. Jin, X. Gao, K. Wong, On capacity of large-scale MIMO multiple access channels with distributed sets of correlated antennas. IEEE J. Sel. Areas Commun. **31**(2), 133–148 (2013)
5. H. Cui, L. Song, B. Jiao, Multi-pair two-way amplify-and-forward relaying with very large number of relay antennas. IEEE Trans. Wirel. Commun. **13**(5), 2636–2645 (2014)
6. L. You, X. Gao, X.G. Xia, N. Ma, Y. Peng, Pilot reuse for massive MIMO transmission over spatially correlated rayleigh fading channels. IEEE Trans. Wirel. Commun. **14**(6), 3352–3366 (2015)
7. S. Jin, X. Liang, K.K. Wong, X. Gao, Q. Zhu, Ergodic rate analysis for multipair massive MIMO two-way relay networks. IEEE Trans. Wirel. Commun. **14**(3), 1480–1491 (2015)
8. A. Adhikary, J. Nam, J.-Y. Ahn, G. Caire, Joint spatial division and multiplexing: the large-scale array regime. IEEE Trans. Inf. Theory **59**(10), 6441–6463 (2013)
9. C. Sun, X. Gao, S. Jin, M. Matthaiou, Z. Ding, C. Xiao, Beam division multiple access transmission for massive MIMO communications. IEEE Trans. Commun. **63**(6), 2170–2184 (2015)
10. A. Liu, V. Lau, Phase only RF precoding for massive MIMO systems with limited RF chains. IEEE Trans. Signal Process. **62**(17), 4505–4515 (2014)
11. D. Kim, G. Lee, Y. Sung, Two-stage beamformer design for massive MIMO downlink by trace quotient formulation. IEEE Trans. Commun. **63**(6), 2200–2211 (2015)
12. Y. Jang, K. Min, S. Park, S. Choi, Spatial resource utilization to maximize uplink spectral efficiency in full-duplex massive MIMO. In *Proc. IEEE Int. Conf. Commun.*, London, UK, 2015, pp. 1583–1588
13. Y. Li, P. Fan, L. Anatolii, L. Liu, On the spectral and energy efficiency of full-duplex small cell wireless systems with massive MIMO. IEEE Trans. Veh. Technol. **66**(3), 2339–2353 (2017)
14. H. Tabassum, A.H. Sakr, E. Hossain, Massive MIMO-enabled wireless backhauls for full-duplex small cells. In *Proc. IEEE Global Commun. Conf.*, San Diego, 2015, pp. 1–6
15. R. Mai, D.H.N. Nguyen, T. Le-Ngoc, Joint MSE-based hybrid precoder and equalizer design for full-duplex massive MIMO systems. In *Proc. IEEE Int. Conf. Commun.*, Kuala Lumpur, Malaysia, 2016, pp. 1–6
16. B. Li, D. Zhu, P. Liang, Small cell in-band wireless backhaul in massive MIMO systems: a cooperation of next-generation techniques. IEEE Trans. Wirel. Commun. **14**(12), 7057–7069 (2015)
17. A. Sabharwal, P. Schniter, D. Guo, D.W. Bliss, S. Rangarajan, R. Wichman, In-band full-duplex wireless: challenges and opportunities. IEEE J. Sel. Areas Commun. **32**(9), 1637–1652 (2014)
18. H.Q. Ngo, H.A. Suraweera, M. Matthaiou, E.G. Larsson, Multipair full-duplex relaying with massive arrays and linear processing. IEEE J. Sel. Areas Commun. **32**(9), 1721–1737 (2014)

19. G. Zheng, Joint beamforming optimization and power control for full duplex MIMO two-way relay channel. IEEE Trans. Signal Process. **63**(3), 555–566 (2015)
20. D. Bharadia, E. McMilin, S. Katti, Full duplex radios. In *Proc ACM Sigcomm*, Hong Kong, China, Aug. 2013, pp. 375–386
21. 3GPP, Universal mobile telecommunications system (UMTS); spatial channel model for multiple input multiple output (MIMO) simulations. In *v.12.0.0, Tech. Rep.* TR 25.996, Jun. 2012. [Online]. Available: www.3gpp.org
22. K.I. Pedersen, P.E. Mogensen, B.H. Fleury, A stochastic model of the temporal and azimuthal dispersion seen at the base station in outdoor propagation environments. IEEE Trans. Veh. Technol. **49**(2), 437–447 (2000)
23. G.B. Giannakis, C. Tepedelenlioglu, Basis expansion models and diversity techniques for blind identification and equalization of time-varying channels. Proc. IEEE **86**(10), 1969–1986 (1998)
24. S. Goyal, P. Liu, S.S. Panwar, R.A. Difazio, R. Yang, E. Bala, Full duplex cellular systems: will doubling interference prevent doubling capacity? IEEE Commun. Mag. **53**(5), 121–127 (2015)
25. E. Everett, A. Sahai, aA. Sabharwal, Passive self-interference suppression for full-duplex infrastructure nodes. IEEE Trans. Wirel. Commun. **13**(2), 680–694 (2014)
26. J. Singh, S. Ramakrishna, On the feasibility of codebook-based beamforming in millimeter wave systems with multiple antenna arrays. IEEE Trans. Wirel. Commun. **14**(5), 2670–2683 (2015)
27. Z.M. Liu, Y.Y. Zhou, A unified framework and sparse Bayesian perspective for direction-of-arrival estimation in the presence of array imperfections. IEEE Trans. Signal Process. **61**(15), 3786–3798 (2013)
28. J.C. Shen, J. Zhang, E. Alsusa, K.B. Letaief, Compressed CSI acquisition in FDD massive MIMO with partial support information. In *Proc. IEEE Int. Conf. Commun.*, London, UK, Jun. 2015, pp. 1459–1464
29. J.O. Smith, Spectral Audio Signal Processing, W3, 2011. [Online]. Available: http://books.w3k.org/
30. S. Vishwanath, N. Jindal, A. Goldsmith, Duality, achievable rates, and sum-rate capacity of Gaussian MIMO broadcast channels. IEEE Trans. Inf. Theory **49**(10), 2658–2668 (2003)
31. A. Barg, D.Y. Nogin, Bounds on packings of spheres in the Grassmann manifold. IEEE Trans. Inf. Theory **48**(9), 2450–2454 (2002)
32. S.M. Kay, *Fundamentals of Statistical Signal Processing: Estimation Theory* (Prentice-Hall, Englewood Cliffs, 1993)
33. G. Caire, N. Jindal, M. Kobayashi, N. Ravindran, Multiuser MIMO achievable rates with downlink training and channel state feedback. IEEE Trans. Inf. Theory **56**(6), 2845–2866 (2010)
34. D. Nguyen, L.-N. Tran, P. Pirinen, M. Latva-aho, Precoding for full duplex multiuser MIMO systems: Spectral and energy efficiency maximization. IEEE Trans. Signal Process. **61**(16), 4038–4050 (2013)
35. B. Hassibi, B.M. Hochwald, How much training is needed in multiple antenna wireless links? IEEE Trans. Inf. Theory **49**(4), 951–963 (2003)
36. 3GPP, Further enhancements to LTE time division duplex (TDD) for downlink-uplink (DL-UL) interference management and traffic adaptation. In *v.11.0.0, Tech. Rep.* 36.828, Jun. 2012. [Online]. Available: www.3gpp.org
37. H.A. Suraweera, I. Krikidis, G. Zheng, C. Yuen, P.J. Smith, Low complexity end-to-end performance optimization in MIMO full-duplex relay systems. IEEE Trans. Wirel. Commun. **13**(2), 913–927 (2014)
38. X. Xia, D. Zhang, K. Xu, W. Ma, Y. Xu, Hardware impairments aware transceiver for full-duplex massive MIMO relaying. IEEE Trans. Signal Process. **63**(24), 6565–6580 (2015)

Kui Xu was born in 1982. He received his B.S. degree in wireless communications and Ph.D. degree in software-defined radio from the PLA University of Science and Technology, Nanjing, China, in 2004 and 2009, respectively. He is currently an associate professor in the College of Communications Engineering, Army Engineering University of PLA. Since 2013, he has been a postdoctoral fellow with the PLA University of Science and Technology. His research interests include broadband wireless communications, signal processing for communications, network coding, and wireless communication networks. He has authored about 50 papers in refereed journals and conference proceedings and holds five patents in China. He is currently serving on the Technical Program Committee of the IEEE WCSP 2014. He received the URSI Young Scientists Award in 2014 and the 2010 Ten Excellent Doctor Degree Dissertation Award of PLAUST. He also serves as the reviewer of the IEEE Transactions on Wireless Communication, the IEEE Transaction Vehicle Technology, the IEEE Communications Letter, and the IEEE Signal Processing Letters.

Xiaochen Xia was born in China in 1987. He received his B.S. degree in electronic science and technology from Tianjin University (TJU) in 2010, and M.S. degree in communication and information system from Army Engineering University of PLA, in 2013. He is currently working toward the Ph.D. degree in Institution of Communications Engineering, Army Engineering University of PLA. His research interests include relaying network, full-duplex communication, network coding, and MIMO techniques. He received the 2013 excellent master degree dissertation award of Jiangsu Province, China.

Yurong Wang was born in China in 1990. She received her B.S. degree in computer science and technology from Beijing Institute of Technology (BIT) in 2013. She is now working toward the Ph.D. degree in Institution of Communications Engineering, Army Engineering University of PLA. Her research interests include full-duplex communication, massive MIMO system, and broadband wireless communications.

Wei Xie received his M.S. degree in communication and information system from PLA University of Science and Technology (PLAUST) in 2002. He is currently a lecturer in the PLAUST. His research interests include wireless communication systems and networks, cooperative communications, and cognitive radio.

Dongmei Zhang was born in 1972. She received her B.S. and M.S. degrees in communication engineering in 1993 and 2005, respectively, and Ph.D. degree in communications and information system from the PLA University of Science and Technology, Nanjing, China, in 2013. She is currently a professor in the College of Communications Engineering, Army Engineering University of PLA, Nanjing, China. Her research interests include new generation wireless mobile communication system, radio resource management, and network coding in wireless communication.

Index

D. N. K. Jayakody et al. (eds.), *5G Enabled Secure Wireless Networks*,
https://doi.org/10.1007/978-3-030-03508-2

S

T

U

Lightning Source UK Ltd.
Milton Keynes UK
UKHW021314080219
336971UK00002B/21/P

9 783030 035075